Praise for the series

'[...]No/Mad/Land examines the challenges
of leadership and the costs of joining a radical
movement for a diverse cast of memorable characters.
Particularly striking are his two charismatic but
flawed protagonists, who accept extraordinary
nanotech biomodifications, only to struggle with
issues of personal identity. This ambitious novel
embraces many diverse cultures with a sensitivity and
insight that will appeal to Francesco Verso's many
international readers.'
James Patrick Kelly, Winner of the Hugo, Nebula
and Locus awards

'The Roamers is an urgent, impassioned work for our
times that cements Francesco Verso's place at the
forefront of European SF writers working today.
Not to be missed!'
Lavie Tidhar, Award-winning author of Osama

'Our world is closing in. We're in danger of
becoming parochial and tribal. This is why Francesco
Verso is such an important voice in SF. Here's a
writer and publisher from outside the anglosphere,
not just reminding us that SFF is a global literature
and a global language but working in tireless support
of writers and literature from all across the planet.
Listen to what he has to say.'
Ian McDonald, Winner of the Hugo, Locus, BSFA
and Philip K. Dick awards

FRANCESCO VERSO

NO/MAD/LAND

Part Two of *The Roamers*

Translated by Sally McCorry

This is a **FLAME TREE PRESS** book

Text copyright © 2024 Francesco Verso
Translation copyright © 2024 Sally McCorry

FLAME TREE PRESS
6 Melbray Mews, London, SW6 3NS, UK
flametreepress.com

US sales, distribution and warehouse:
Simon & Schuster
simonandschuster.biz

UK distribution and warehouse:
Hachette UK Distribution
hukdcustomerservice@hachette.co.uk

Publisher's Note: This is a work of fiction. Names, characters, places, and
incidents are a product of the author's imagination. Locales and public names
are sometimes used for atmospheric purposes. Any resemblance to actual
people, living or dead, or to businesses, companies, events, institutions, or
locales is completely coincidental.

Thanks to the Flame Tree Press team.

The cover is created by Flame Tree Studio with
thanks to Shutterstock.com.
The font families used are Avenir and Bembo.

Flame Tree Press is an imprint of Flame Tree Publishing Ltd
flametreepublishing.com

A copy of the CIP data for this book is available from the British Library
and the Library of Congress.

1 3 5 7 9 8 6 4 2

HB ISBN: 978-1-78758-927-8
PB ISBN: 978-1-78758-926-1
ebook ISBN: 978-1-78758-928-5

Printed and bound in Great Britain by Clays Ltd, Elcograf S.p.A.

PART TWO

NO/MAD/LAND

'I wouldn't be surprised if people reading this story don't believe what
I am referring to, especially people who have never travelled, because
people who haven't seen much don't believe much, whereas people
who have seen a lot believe more.'
Fernand Mendez Pinto, *Voyages and Adventures*

The reason why humans moved on from being hunter-gatherers to
farmers is one of the most ancient and disputed questions in the history
of humanity. Many scientists believe this step was one of humanity's
worst mistakes.

In general, becoming sedentary enabled people to spend more
time on artistic activities and developing new abilities, at the same
time freeing them from the anxiety of daily nutrition. However,
even though agriculture renders more per unit of land (a community
of twenty-five people can feed itself with twenty-five acres of land,
whereas with the hunter-gatherer model it would take thousands) the
same cannot be said of the quality of food produced per unit of time.
In other words, agriculture provides a diet with less variety in it than
that of a hunter-gatherer community. For example, the Aboriginals
eat seventy-five types of wild plants, rather than only depending on a
few cultivated ones.

It is only in modern times through the global market that humanity has regained the possibility of having a more varied diet, but this is only true of the richer areas of the world and for a small part of the population. On the other hand, nomad societies have been present throughout human history. It is even possible to see how this phenomenon grows and reaches new heights every time an anthropological paradigm reaches a point of crisis, every time an 'old world' dies to make room for a new one. Thus, in the West, as with the start of the second millennium, or during the Renaissance, there is a proliferation of millenarian movements, mystical ferment, and religious disorder of all types. In each of these cases there is a manifestation of progressive saturation of popular imagination and, before a new legend can form and supplant the preceding one, the customs and behavior of the masses tend to wander for a while, following alternative routes, along with an increasingly insistent search for new experiences: a sequence of – more or less improvised – attempts and fatal errors, which despite everything turn out to be fundamental to guaranteeing a defined shape for future social structure.

It is during these crucial periods that the theme of escaping from what is considered a decadent world seems destined to reacquire sense and meaning. Because, even if it is only over the timespan of a generation, what exists stops being enough and the values of a specific period are no longer able to satisfy new expectations. Social revolt, like small acts of daily rebellion and provocation, becomes uncontrollable, and trust in principles shared until a short time before suddenly vanishes: from this moment on society no longer retains the same self-awareness.

This phenomenon can also be observed at the start of the third millennium. In different forms: music, cinema, literature, art, and even simple conversations tend to express and transmit a level of disgust in a world known by referring – in ever greater levels and pervasively – to artificial paradises, hedonistic illusions, escapism and otherworldly solutions.

Decadence is frequently a symptom of a metamorphosis or transfiguration in progress, both on an individual and a social scale. We can complain about this foment, we can deny these tendencies, but

ALAN CORMANI

CHAPTER TWENTY-NINE

Transitions

The path in the middle of the bush is quiet except for the sound of footsteps. His heartbeat and breathing are not accelerated, nor is he sweating. Alan watches the moving clouds in the sky. Step after step, feet slapping the ground at a regular pace encourages his thought processes: every rhythm starts with the feet.

Lowering his gaze, he sees the first kind of writing man ever learned to read, footprints: signs impressed in the earth that enable whoever knows how to interpret them to evolve, as individuals and then as a species.

The footprints don't belong to dangerous animals, nor prey, but the six palomino horses, four dingoes, eight merino sheep, five Maremma cows, and two Barbary apes Rome's zoo didn't want. They are tame animals and as such wouldn't last a month in the wild. If anything, danger was an intrinsic part of their route: the Pulldogs are heading north, to the Mount Adone Wildlife shelter in the Tuscan-Emilian Apennines.

The phone in his trouser pocket is ringing: this is the second warning.

Alan has taken one hundred and seventy thousand steps, three times what he normally walked in a month before his organism's internal revolution. He's worried though, filled with a prickly anxiety that keeps nipping into his muscles every time they go over a hill and there is neither a river nor a spring anywhere to be seen.

"What's our situation, Dikran?"

"Negative," he answers after checking an app on his phone. "This area used to be full of streams, rivers, lakes, and springs with every hour of walking."

"What happened?"

"Privatization. Every drop of water has been diverted, channeled and put underground.... At best, the rest has dried up or is being fed to the factory farms in the surrounding valleys."

Alan checks his water bottle. Empty. His old neurons, the ones that haven't yet gone to make room for their copies enriched with nanites, tend to take that for granted. How long will it be before he stops thinking in terms of taps rather than springs and fountains?

His phone tells him his temperature is thirty-eight-point-eight degrees Celsius, but he doesn't have a fever. He ups his pace, overtaking several of his comrades to reach the front of the group. In about thirty thousand steps he will have to work out a way to cool down his and the others' bodies, especially the children.

The human body can survive without water for a week, ten days at the most. However, humidity, temperature, physical activity, and the size of mineral and fat reserves can reduce this safety period to four; in Alan's mutated condition it's even less, three days: seventy-two hours of autonomy before he begins to feel unwell. This is because – except for Rafabel and Hakim – no one else accepted Nicolas's proposal to increase the efficiency of their nanites through artificial photosynthesis. Their skin overheats rapidly and they need even more water to avoid becoming feverish, which would force them to slow down, if not stop altogether to lower their internal entropy.

At the top of the hill Alan raises an arm to stop the group. He turns and realizes he hasn't heard the noise of an engine for several days. The only engines around are airplanes flying at cruising height over Mount Amiata, innocuous for eyes and ears. Far from the line of the A1 Rome-Milan motorway, Alan and the Pulldogs have entered another century: the silence surrounding the Tuscan hilltops is the expression of an earlier era, a bubble of existence where high-tension cables and electronic devices are rare.

The sound of nanites splitting atomic bonds, making proteins, vitamins, and carbohydrates, transforming tubers, leaves, mushrooms, and forest fruit into nourishment in their bodies is infinitely lower than the buzz of a beehive going about its daily business.

The only sounds accompanying their walk are the singing and chirping of birds, the neighing of horses, and the yowling of skinny dogs alarmed by the passing of strangers.

Despite this, Alan is afraid that sooner or later one of the Pulldogs will get bored, or scared, and reject this sudden change and go back to Rome. Over the last few nights they have been camping near the Park of Monsters in Bomarzo and have traveled along the Etruscan Vie Cave from Pitigliano to Sovana, walking along the incredible sunken lanes excavated in rock. But melancholy is an insidious thing that can affect anyone, it can initiate a dangerous sense of nostalgia, especially in special cases such as seeing an empty pool in the grounds of a deserted mansion, sitting around the tables of an old restaurant that specialized in serving wild game, or faced with the simple view of a service station, signs all lit up to attract drivers as if they were moths. For Alan, though, these first few days of walking have been the perfect occasion to find himself, strengthen neglected relationships, and make difficult decisions.

Coming out of his walker's trance, Alan realizes Silvia isn't in front of him. Since leaving Rome, her mood has lightened and a few times, influenced by the atmosphere of a starlit night so cool as to make her shiver, she has come to him, in a way that she hadn't for a while. When she dug her nails into his back, her thighs tight around his waist, he felt her energy again, the same electrifying strength that had run between them during the first years after founding the community on the Garbatella-Testaccio viaduct.

He looks for her; she is next to Rafabel, who is telling little Tasia and Pino legends about the trees they pass along their path.

"Auntie Rafabel? I'm thirsty..." the little girl says.

Rafabel's side parting and the smattering of freckles on her face make her look younger than her forty-five years. She pulls a couple of hazelnuts from the pocket of her ethnic-patterned dress and offers them to the kids.

"Suck these, it will stop you being thirsty."

Close by, Kenshij nods, his gray ponytail bobbing up and down, confirming her advice. The man doesn't look at where he's putting his feet. He is barefoot and touches the ground as little as possible, moving forward with great elegance. He hardly leaves a sign of his passing, gliding rather than walking. He rarely talks, he simply acts and when he's in a chatty mood he mutters either yes or no depending on the situation.

Behind him Leira's and Ariel's silvery dreadlocks bounce in unison. The Rastafarian twins drum their fingers on bongos hanging from their belts, while Dikran checks his phone's screen and the topography of the land, hunting for their next stop. At the back of the group Miriam is riding the young palomino mare, Nanà, while Martina and Valeria are leading the other horses by their halters. A little way off Tommaso and Pilar – newly become lovers – hang back, exchanging tender words.

Suddenly Dikran speeds up to walk beside Alan and points out the valley opening to their right. "Water!"

In the distance, a yellowish smear surrounds a petrol station, with a bar.

"Yes, but in bottles," Alan replies, annoyed, "and we don't have any money."

The neon sign says: *The Shooting Star Bar*

"I have an idea," says Valeria.

Inside there are two elderly customers and a bearded man behind the bar drying glasses.

"We've followed the shooting star and it brought us here," starts Valeria happily.

"Welcome..." the barman answers, joining in the game. "Unfortunately, the star passed by a long time ago. You mustn't expect too much from this eatery. Since they built the bypass no one comes by here anymore. What can I get for you?"

"Water," says Alan.

Valeria touches his arm, reminding him to let her do the talking.

"We'd like some crates of water…for our journey."

"Where are you going?"

"North, to the Tuscan-Emilian Apennines."

"How many crates?"

"As many as you have."

He doesn't miss a beat but he moistens his lips as if evaluating the strangers. "I'll go and see what I have in the storeroom."

The elderly couple leave a crust of bread on their plate and stand; the man leaves a banknote on the table and follows the woman as she leaves.

"We're going now, Alarico."

The barman shouts out a goodbye and comes back carrying two crates of water.

"I have these…and another three under the bar."

"That's plenty, thanks."

"Thirty euros please."

Valeria unleashes her trap, her smile. "If you accept the gift of the three wise men, this place will really become a shooting star."

He frowns and waits to hear the details of her proposal.

"The sign outside…. We can turn it into a shooting star. I'm not kidding, your bar will become the main attraction of the valley."

Alarico bursts out laughing. "You're a funny lot, you are…. Let's see what you can do."

Two hours later the bar's sign has become an installation: every time a shooting star or a fragment of orbiting rubbish crosses the sky above the petrol station, a beam of light will follow its trajectory until it vanishes. In an hour, under a clear sky, it's possible to see two or three pass by, instant flashes of light coming and going within a few moments. Even when the sky is overcast, the lasers will light up colored trajectories slicing through the clouds of the night sky.

The Pulldogs sit in a circle on the side of a wooded hill to watch the spectacle, handing out water to the thirsty.

"I wrote the program for the Cimini Astronomical Observatory," says Valeria, taking a sip of water. "It takes data from the Italian Space

Agency and translates it into real time. The lights are simple LED. For the rest I only needed a laptop and the projector Alarico uses for watching football with his friends."

"It looks like that game where you had to shoot down alien space ships..." says Alan.

"*Space Invaders,*" answers Nicolas.

"Now we can set up camp. Will down there be all right?"

Nicolas is bobbing his head up and down. The music he's listening to through headphones slides over him, causing dynamic pictograms to form down his shoulders and along his arms. His variegated pigmentation is the result of modifiable chromophores. The bathochromes react to the bass notes, turning his skin a reddish color; the hypsochromes capture the higher frequencies and turn light blue.

He drops to his knees and analyzes the ground. He uses his eyes, and then his fingers because his nose hasn't been much use since his father, Pietro, punished his decision to leave the family perfumery. When he opened the letter from his father ordering him not to use the smartfumes of his own design, he was hit with a permanent anosmia caused by something also contained in the envelope. Only a small whiff of it caused Nicolas to lose his sense of smell.

"Excellent choice. We have the molecules we need."

The wide bandolier he's wearing carries a number of colored vials. Inside these are substances (aluminum, carbon, chalk, and cellulose powders) useful in the composition of any basic necessities. The other Pulldogs jokingly call this his paint palette, but Nicolas uses them to create materials, not to paint.

The place they have chosen for the day's nest is a cavern in a spinney of chestnuts. The trees are scaly like reptiles and give off a strong perfume of resin. The thick knobbly branches, covered in green and silvery lichens, look to them like shadowy faces and fantastic bodies like the stone monsters they admired a few evenings ago in Bomarzo.

Behind them they can see the silhouette of Mount Amiata, and not far off is the ancient line of the Via Francigena, now followed by Sunday cyclists, seasonal pilgrims, and hikers hungry for a taste of the country.

Though the cyclists aren't a problem, they greet the Pulldogs with friendly waves and pedal off quickly. They have to keep everyone else at a distance in case they ask too many questions. (What's with the animals? Where do you sleep? What do you eat?) They could get suspicious and report them to the police. Then any exchange of seasonal produce would be worth nothing, the sight of the children or their reason for taking the animals to the Mount Adone center, though true, wouldn't be enough to guarantee an easy passage.

It only takes them a few minutes to set up their tents. Their portable nanomats – like agile biomechanical spiders – start weaving arched filaments, making jointed sticks like canes and covering surfaces, following compositional instructions shared to everyone's devices by Nicolas and using the reserves of raw materials that had been replenished on-site. The configuration of each tent can be saved and reproduced, exactly the same, day after day, or changed depending on the materials available to them, according to taste and preferences, and the month or the season.

Up until now the tents have ensured they spent their nights at a reasonable temperature, blend in with the forest, and protect themselves from bad weather, but if needs be they will also be protection from wild animals, intolerant farmers, and forestry police who might mistake them for gypsies or fleeing immigrants. When it needs to, the nanotech material the tents are made of can harden to form an impenetrable shell.

As soon as Nicolas's circular yurt is finished, he begins to gather twigs. Then he snaps them and puts them in a basket. "We should have gone south," he mutters to himself. "We should have got on the first ship and headed towards the Mediterranean."

Instead of waiting for evening and talking about it all together around the fire, over the last few days Nicolas has begun to moan and whine at regular intervals. Sitting on a fallen trunk, Alan is waiting for his nanomat to finish building a bubble that kind of resembles an igloo. Every now and then he plucks a string of his *guitortoise*, a musical instrument made using a tortoise shell.

"You're following a track that will take you nowhere, Nico. I agree with the general idea, but think about it, we can't go down there, not now. We have women and children and we have to deliver the animals. We aren't ready, the whole Mediterranean area is dangerous. As your friend, and a sane human being, I have to warn you that things might not go as you think they will."

A tall, imposing figure draped in a colorful tunic is following Nicolas and helping him pick up twigs. The proximity of Hakim to Nicolas isn't only physical; he communicates his agreement or not to a proposal through his vicinity.

For Alan, Hakim has always been a mysterious subject. Sometimes he finds Hakim walking beside him and they exchange a handful of words. It was the same when they were still on the viaduct, he would come back late from his shift at the zoo and leave early, and at other times he would say few or no words. Alan generally doesn't need a lot of chat to get an idea of a person. He might be wrong sometimes and make a hurried judgment, like with Nicolas, whose potential he underestimated, but with Hakim he believes he has got it right, even though the other Pulldogs have varying ideas: Miriam likes him, maybe because he takes care of the animals, whereas Silvia sees him as too introverted and reserved.

Two days earlier, during supper, they had heard a loud, sharp noise coming from the forest. Perhaps a twig snapping or a falling pine cone, making Hakim jump up like an animal in alert mode. Silvia had looked at him scornfully, as if to say he was exaggerating with his superhero display.

Once, when Alan had moved a little way off in the middle of the night to empty his intestines, he found Hakim awake and keeping guard against who knew what. When Alan asked him why he wasn't sleeping he replied, "The night is full of predators."

After leaving the twigs in front of his yurt, Nicolas heads towards the stream to collect a little water. When he gets back to the clearing, he picks up two cartridges, pours a few drops into the nanomat and makes an infusion.

"Going south is a good idea, to democratize the nanites," he says. "I've already got a slogan, 'nanites for everyone'."

In the meantime, Hakim has lit the fire and is putting water over it to boil. Nicolas is still holding a chestnut leaf as he continues to expound his reasoning to Alan.

"We have to spread our way of living, we have to free people from the need for food."

"Yeah, right, like the other day," Alan says sarcastically. "When you suggested it to that farmer who liked the idea so much he shot at us...."

Nicolas rolls with the blow and starts waving the leaf around, then he starts stroking it, tracing its veins with his fingers.

Rafabel comes to his aid. "He's right, Alan.... Plant cells send chemical signals like the ones neurons send and know how to evaluate the world and make decisions accordingly. This should convince us to look at every vegetable, including weeds, under a different light."

Nicolas scrabbles around in the earth, looking for something. Then he holds a mushroom up to his nose. Disappointed by the uselessness of the gesture, he passes it on to Rafabel. She sniffs, decides the mushroom is edible and says, "Personally I admire myxomycetes, appreciate lichens, fear aconite and respect the castor oil plant."

"We don't want to preach," Nicolas interrupts. "We only want to show you how life nests in all of its hosts, but in different ways: we gather knowledge from the outside, from books, external memories, the cloud, and this choice put us on a learning curve no other species has access to. Sadly, we have abused this advantage, we have drained nature, and now there is nothing left for us to do but learn to control our predatory instincts and give other species some knowledge back in the form of nanites."

Alan doesn't seem impressed. He stops plucking at the strings of his guitortoise and, putting the instrument over his shoulder, jumps off the trunk.

"That's as may be, but look at us...a handful of grains of sand has never started an avalanche. You were there too when we decided to head north. First we have to make sure the animals arrive safe and sound at Monte Adone, then we can reconsider."

The water is beginning to bubble. Nicolas stands and puts some cups out on a flat log.

"I agree about the animals, we can't take them with us. But Alan, the multitude makes the difference…. A network with millions of joints behaves differently to one with hundreds, like an ant colony compared to just a few ants. Adding walkers to our group will increase the value of each of us."

Alan shrugs and is about to walk away when Nicolas shouts out, "Tea's ready! Come and get it!"

Nicolas hands Alan a cup as if it were the pipe of peace. Then, about two dozen thirsty people arrive.

A dot shows up on the screen, about three kilometers from the camp. It moves forward. Stops. Goes back on itself. Stops again.

As he sips his tea Alan checks Rafabel's position on his phone; she and Silvia went for a walk a little earlier. Knowing that Rafabel is scared of getting lost and has a terrible sense of direction, he had given her a GPS bracelet so he could always find her again. It's a minimum condition to feel safe wherever she goes.

She and Silvia will have already unleashed swarms of nanites in the Piancastagnaio territory and now – having tracked the pollen and perfume drifts using the omnitrack app – they will be on the verge of discovering where the best flowers grow and the tastiest berries and juiciest fruits are hiding to gather for supper. The Pulldogs' diet, though sporadic and consisting of one meal every three or four days, is more varied than it ever was in Rome. Each territory is like a different restaurant.

Rafabel and Silvia's 'shopping list' includes jujubes and chestnuts for roasting, Jerusalem artichokes, blackthorn berries, porcini mushrooms, hazelnuts, and acorns. That is not all; they are also looking for blackberries, raspberries, and blueberries in the forest undergrowth, and honey and royal jelly from beehives. Winter has been demoted from general to lieutenant, making fruit and vegetables available that up until a few years before would have been out of season.

After filling a few baskets Silvia and Rafabel decorate their delicacies with cyclamens, holly, and broom; a pair of green pine cones, a few twigs of spruce, and finally they are ready to load them onto the hover transporters, loyal solar-powered stretchers fitted with proximity chips, waiting to carry everything back to camp.

When Alan sees them handing out food to the others, the argument is still going on.

"...just suppose for a moment that our current civilization were destroyed," says Nicolas.

"Why should that happen?"

"Because the wheel has had its time, money has had its time, and life in cities is not the pinnacle of satisfaction."

Nicolas takes his plate and place in line with Alan following him.

"Those things will never vanish. There will always be someone who can pull them out again, they will open a history book, read a book of instructions, download a town plan, and re-propose the same work-in-exchange-for-money and safety-in-exchange-for-happiness model."

"It wouldn't be so easy.... Trust me, Alan, things never go back to how they were. The prehistoric wheel is not a pneumatic tire; even if people found a manual it would take a long time before they could understand much, or maybe they would have to find someone who could read and understand the instructions. Once those people are dead, they will have no more apprentices, there will be no successors."

At the word *successors,* Nicolas's face darkens. Leaving the Rendezvous has had deep consequences: an irreversible choice that has burned him, the result of which follows him even hundreds of kilometers away, every time he tries to sniff a flower. His acquired lack of smell has excluded him from taking part in food gathering, and when the others go to explore the wooded slopes or the wild meadows, he stays in the camp making and dismantling what they need, as sad as a sickly child who can't join in with the games. Even when Rafabel drags him along with her to stop him getting too depressed, he moves clumsily because even though he can touch and see, he gets no precise olfactory information from what he is collecting.

"One generation would not be enough. There is only one way to disperse so much knowledge in such a short time: revolution."

"Revolution, Alan? Revolutions have never been good for the people. I say it would be enough to break the link between one generation and the next."

Even though he is talking about society in general, Alan can see he is referring to Pietro. From what Nicolas has told him, without his ability to assemble molecules, his father would not have been able to create a perfume worthy of the name. Actually, if he didn't make up his mind to hire a new designer, the business would fail within the year.

When he looks at Nico, Alan cannot believe the change that has come over him: when they first met, during a game of Urban Golf, he had simply been yet another daddy's boy, rich, clumsy, and insecure, oppressed by the weight of his family's expectations and unable to make decisions for himself, whereas now he seems so at ease in his new super-heated skin and this unknown world continually changing under their feet.

"I don't know, Nico. You're talking about a lot of abstract stuff. I am already walking to leave history behind me. Day after day, all this surrounding us is becoming our only reality and all the rest is just nonsense, written on the edge of a map that no longer has any relation to what is out there."

"I love maps. When I was small, I drew countries, I knew all the capitals by heart, I used to trace the globe with my fingers, dreaming of adventurous voyages."

Alan turns and looks for Silvia amidst the tents. She is still sorting out the baskets, so he carries on talking. "I want to leave too. Let's do what we have to do and then go, beyond Europe, towards the Baltic states. Then we can lose ourselves in the Russian taiga."

Miriam comes forward, sipping her tea. "That would make Ivan happy.... I'm sure he is waiting for us. Who knows what he's doing? That man can never stay still."

After getting out of prison, where he had been for violating the Ending Hunger project's copyright, old Ivan Shumalin went back to his country, to a small town called Priozersk on the shores of Lake Ladoga in the Russian-speaking Karelia.

"That's exactly who I was thinking about. Last month he told me he's working on a project involving distributed mass computing, the Green Ark, connected to the international network of the Micronations. I want to go see him really badly."

"We'll go sooner or later, son. I miss Ivan a lot...."

Alan stands and finishes his drink. "I am convinced there are two worlds now: one of them is not ours. They run together in parallel, without either understanding or accepting the other."

Nicolas nods.

There are some noises on their right, shouts and screams of excitement echoing through the trees. Pilar and Tommaso, Tasia and Pino are splashing around in the stream of water and carry on doing so until the burning sun forces them to look for the cool of some shade. Then they will all grab handfuls of mud and spread it over their shoulders to help dissipate the heat of their bodies. Towards evening they will explore the maze of paths and tunnels of wild undergrowth, amongst curved roots, creepers, and leaves hanging down like the ropes in an adventure playground. Then finally, before going to bed, they will play music and sing under the light of the moon reflected in the water.

At the nest the children are running around like baby chickens in the meadow and wandering through the trees looking for something interesting to peck, an insect to play with, a flower to sniff, or an acorn to kick and chase.

Alan goes to fill his cup again. Not for himself though. This time he takes the drink to Silvia. When she realizes he's near her she tenderly pulls his ruffled beard and kisses him on the lips. Then, sniffing the smell of the tea composed by Nicolas, pushes the cup away.

"Thanks, but not now."

"Aren't you thirsty?"

"Yes, but an infusion of oak and ash? I don't like it. Nico can't smell anymore."

"Are you sure that's why? I think things are working out fine. Even after walking for weeks, no one is showing signs of giving up. As for Nico, he moans and whines but in the end he adapts."

Silvia finishes handing fruit out to the children and leans against a tree.

"It's early days yet. For many of them this adventure is a kind of holiday in the woods, away from the city. Think about the adults, any jobs they had before would have infuriated them long before this."

"To be honest, I'm worried."

"You? Worried? Look up. You see the sun's rays coming through the branches? They have come a hundred and fifty million kilometers. Now that is a journey."

Alan pushes her gently against the tree trunk. Their relationship, like the weather, is a chaotic system ruled by forces that are too unstable and complex to forecast. However, certain behavior, over time, cannot be explained by simple chance.

"D'you know what I want? I'd like it if my footprints were not marks in the ground but a trail of musical notes and melodies, like the Songlines of the Australian Aborigines. Bruce Chatwin said, 'A songline was both a map and a direction finder. Providing you knew the song, you could always find your way across country.'"

She moves and takes his hand and walks to the bank of the stream, where the kids are still splashing around.

"Admit it, Alan…. This is a holiday for you too. You were wearing the boss's uniform but behaving like a rock star between concerts. You didn't care about anything, in pure anarchic-Romanesque style. Then you realized you had to make plans. Moving around, having to choose a daily destination, have forced you to think of a route, places to stop and rest. That's something. Desire reenforces the attitude, but I would hate for you to end up feeling like a big fish in a small pond."

"A rock star?"

"Yeah…. One day you'll organize the world tour."

She smiles as she says it but as soon as she stops speaking she turns away, bending over and holding her stomach.

"Are you all right, love?"

Another spasm makes her bend over again. And again. The third heave leaves a pool of vomit on the ground. It is dark and metallic looking.

"There we are," she says, pointing at it, "forty years of medicine with mercury in it. The nanites don't like it at all."

Alan looks closer and sees there is enough mercury there to fill a glass half full. Then he notices the dark marks on Silvia's skin.

"Are you ill? You haven't taken Nicolas's heliotrons, have you?"

"No, this is the poison I've been carrying around with me since I was a kid. When your body expels poisons, it does it through the skin too."

"But so much? Nothing like this happened to me."

"My mother was obsessed with medicines. It helped her anxiety. Our house was like a chemist's. Who knows how much she used to sneak into my breakfast without me knowing."

"So those marks are a reaction to the nanites?"

"According to Nico, those splodges are the reaction *of* the nanites. Call it antivirus if you prefer. He calls it a 'remote immune system' because the nanites are capable of downloading health updates from the web."

"Well, that means you are detoxing," Alan says with irony.

A slight smile crosses Silvia's face, though her eyes are full of tears of pain. She rips a large leaf from a bush and wipes her mouth.

CHAPTER THIRTY

Fosso Bianco

The bushes become forest when the Pulldogs push through a thick wall of beech trees like needles to walk on a mass of leaves where human feet have never touched. There are no footpaths here and although there must be one somewhere, it seems like any trace of a path has been covered by the juniper bushes, brambles, dry twigs, and crumbling sandstone. It could be years since man last set foot here. It doesn't take much to erase evidence of man's presence from the land: the forest rapidly avenges every trespass by incorporating even what man has built to last.

Above the carpet of leaves each plant seems different from the next. Of the fallen trees, some are beginning to decompose, and from their rotten trunks the tender shoots of other plants are beginning to grow.

Alan is surprised. Walking deeper into the forest with a mixture of admiration and uneasiness, he brushes the bark with his fingers and breathes in lungfuls of a smell that invigorates him. He listens to the gloomy and excited songs of the birds he doesn't recognize and realizes what he's used to is only a pale copy of true wild nature. The most extraordinary thing, though, is the sensation of familiarity he can feel on a cellular level.

With every step the Pulldogs take further into the nature reserve, the singing and chirping of the birds gets louder and louder. At a certain point the whole cliffside following the southeastern side of Mount Amiata begins to sing.

Hakim is the only one who can identify their individual songs. It is so beautiful it can be described in two overflowing sentences:

"It's like the aviary in a zoo. The air smells of humus like in a greenhouse."

Then what seems like thousands of birds fly out of the hundreds of tunnels and nooks and crannies in the cliff. Alan motions to the Pulldogs to stay still. The group becomes immobile. Then he lowers his hand and gestures them to sit on the ground. He brings a finger to his mouth and then to his ear as if to say, "Close your eyes and listen."

On first impact the cacophony seems a little threatening, but slowly becomes a festive, raucous song.

Unable to resist his curiosity, Alan opens his eyes and sees them all perched on a cliff at least ten meters high. It is a breathtaking spectacle. Slowly he points his phone and sees hundreds of canaries, green- and blue-headed parrots, robins, pipits, and white-eyed parakeets.

The tuff cliff is a changing rainbow, symphonic and deafening. When entire rows of perching birds all sing at the same time like they were a choir, it is as if the branches rustle in the breeze they make, but listening more carefully shows the united sound separates into a knot of hundreds of soloists, the noisy twittering of a multitude of individuals.

Suddenly two flocks rise into the sky; it looks like they are arguing, pushing against each other, diving sharply to demonstrate some right or the other, claiming a better place or just more attention.

"What do you think they are saying?" Silvia says, resting her chin on Alan's shoulder.

"They are friends. I think they are talking about where to go and eat later. They are arranging to meet before breakfast."

Tasia can't resist taking a seed from her pocket. She's immediately surrounded and covered by a cloud of small green parrots, happy to take advantage of a free meal.

The show lasts for a few minutes, then each bird goes its own way, flying alone or in a group through the forest. Alan moves to restart their walk. When he stands up he realizes that many of the Pulldogs kept their eyes open and hadn't lost the opportunity to record the natural concert.

"Marvelous! We've been to our first bird rave party."

In the evening, after the shadow of Mount Amiata has come down like a sheet over a corner of the forest, Alan falls in beside Nico.

"Have you seen the marks on Silvia's skin?"

He nods.

"And the vomiting?"

Another nod.

"Do you think it's serious?"

The group is skirting around the ruins of an old monastery. The building is majestic, a rectangular cloister adorned with columns and an ancient wisteria plant climbing to the top of a crumbling tower.

"The nanites are programmed to react to the environment surrounding them. Neither more nor less than what genes do."

A chapel in limestone with pointed arches, though abandoned, retains its dignity. Peering through one of the windows in the wall they can see sodden mattresses and tinned tomato crates piled up along the walls, two confessionals being used as wardrobes and some dilapidated chairs.

"Silvia vomited two hundred grams of mercury. She said she had purged forty years of medicines. Does that seem like a good reaction to their environment to you?"

The bright green meadow below the apse has won the battle against the flagstones, and the frescoed plaster has succumbed to damp. What must once have been religious magazines for the faithful have been reduced to piles of stained cellulose.

"It depends, because as they work faster than genes, nanites' reactions seem prodigious." Nico points at Alan's legs. "You should know better than anyone else."

Alan hasn't forgotten the accident at Globalzon, nor what Miriam and Ivan did to give him back the use of his legs. He can remember his old organism, his heart pumping blood independently of his desires, lungs breathing even when he was sleeping, neurons activating with every thought, a heritage of unconscious functions given by nature to all living beings...but his bone marrow was unable to repair the connectivity of his Schwann cells damaged by a fall in the warehouse.

A human being's central nervous system does not stretch, not even a little, and if it snaps – unlike that of many animals – it stays inert, like Alan would have done, as a paraplegic, if it hadn't been for the nanites. After swallowing them though, his 'neuroms' stretched and within a few months he was able to walk again, better than before. In any case, he has never felt like a self-aware android worrying over the question: 'Who programmed me and why do I function in this way?'

The creator of his nanites has a name and surname, even if neither Miriam nor Alan have ever met them. They know this person is Chinese, has been living in Italy for years, and gave this nanotechnology knowledge to free distributed cooperation projects. In his hands a generic assemblage engine, thought up by Ivan Shumalin for the Ending Hunger project, had made a miracle possible.

"You know very well that those nanites had a different purpose. They were supposed to repair my damaged bone marrow, not improve the efficiency of how we nourish ourselves. I'd like to know if there will be other side effects that will transform our lives into nightmares."

"Would you have felt better if Silvia's organism had expelled a milligram of mercury per month, Alan? I mean, we know what is happening to the outside of our bodies, but inside? Many things are a mystery, with or without nanites. Is the intelligence working in cellular mechanisms so different to that regulating nanites' microscopic processes?"

"Nico, I asked you a simple question and you answer me with three other questions. I just want to know if Silvia might have the same illness as that bastard Grisha."

"He wasn't as ill as he wanted us to believe. He had an endogenous fever like all of us, perhaps a little higher, but in the end he decided to leave the viaduct and betray us for thirty pieces of silver."

"So it isn't the same thing?"

"I don't know. Nanites do good and bad. They are like the wind, they blow in every direction."

"That's a shitty answer."

"Do you want another one? Here you go: I'm not a doctor, nor do I think there is anyone specialized in medicinal nanotechnology, not yet at least."

Alan says nothing, but continues to walk by Nicolas's side.

There are many points where the stones have fallen, taking large parts of the facade with them. Inside, something is making noises, there are probably owls or bats living in the bell tower or the cellars. The floors of the second story are only held up by rusty arm-thick metal struts and Rafabel has to make sure Tasia, Pino, and the smaller children don't go and play underneath.

Alan walks on in a state of agitation. Every now and then, almost as if trying to attract Nicolas's attention, he kicks at bricks in the fallen perimeter wall. Then he grabs a crumbling stone and crushes it between his fingers.

"What a place…. Time has won here. There is no future here."

"I can't even see a present here. Tell me how it smells, Alan."

"It stinks…it's disgusting."

"I thought so. Let's not set up camp here tonight. Shall we go back?"

"Back where?"

"I saw a footpath, three hundred meters before we got to the monastery. If they built it here the monks must have had access to a freshwater source."

"Perhaps you're right, on the other side of the Val d'Orcia I saw two modern buildings that look like fairly new hotels." Alan checks his smartphone. "Found 'em. It is the San Filippo baths. We can look for a place there."

"OK, but if there are hotels nearby, we'll have to be careful."

Alan leans out above a really steep slope. "Hurry up! It's incredible!" he calls out.

The others run to him and spread out beside him.

"Be careful, there's a lot of mud and a layer of slippery leaves too."

Then he starts to make his way down and is the first to get to the bottom, waving his arms enthusiastically.

"What is it?" shouts Silvia, only halfway down.

The slope takes them to a white hill made of limestone formations. Alan climbs up to a plateau to stand hands on hips with his feet in a pool of naturally sulfurous water.

"Our swimming pool. What do you think?"

Fosso Bianco is a pleasant discovery: along the river fed by various sulfur water springs there are a number of natural limestone pools where the younger members of the group are already splashing around. In the woods Rafabel is gathering silver-shelled snails as if they were gemstones.

"Are you sure this is a public place?" Miriam asks her son.

"Well, I didn't see any gates or no-entry signs."

The Pulldogs' skin is impregnated with the perfume of lemons. It is a feature that Nicolas wanted because mosquitoes hate this smell and it keeps them away. This is especially useful when the weather starts to get warmer again.

The air is damp beneath the fir wood's canopy and the high clouds look like sculpted marble. Sitting in a circle around the fire, the Pulldogs are doing something unusual for them.

There are evenings like this when they discuss the importance of walking and the value of a volatile community, kept together by steps that are renewed every day rather than by rarely questioned social ties. Other times it is their daily difficulties that make their union desirable, like when they are chased away by a round of blank shots, or when they are walking in the pouring rain. Even though it is January, it hasn't snowed.

When evening comes down, the day never seems to have passed like a normal day, nor is it broken down into hours, minutes, and seconds, ground into pieces by the flowing of digital time. The only ritual that marks the passing of time is the setting up of the communal tent. Once a week, instead of putting up single, double, or family tents, they mount a structure capable of welcoming the whole group. Sleeping one next to the other guarantees a sense of deep togetherness, the kids feel safer, the young people have fun, and the adults socialize.

Usually, during these evenings everyone sleeps better, but tonight, shortly after one, the blue lights of a forestry police car are reflected in the puddles in the road leading to the San Filippo baths. Ariel and Leira are the lookouts, and as soon as they see a few shadows holding flashlights they alert the others with a bird call.

After leaving their car by the side of the road three policemen continue on foot along the path down to where they are camped.

The bird call alarm travels quickly and when the policemen reach the camp they find a group of sleepy, but standing, people and various animals.

Rafabel shoots a quick look at Alan, who immediately starts talking.

"Please, we just need some water. We couldn't find any anywhere else around here. We won't stay long, we'll leave before dawn."

The largest officer, gray moustache and crew cut, ignores Alan's proposal and turns his radio on.

"There are about twenty of them," he says to someone on the other end of the radio. "A couple of old people, some children and some animals, horse, sheep, cows.... Should I move them on?"

Officer Tash nods and waves at the other two to follow the order.

Seeing as the soft approach gained them nothing, Nicolas tries another way.

"We aren't doing anything wrong. This is a public area."

The two officers circle Nicolas. They point their flashlights at his feet then bring the beams upwards, to his torso, and then his face, irritatingly in his face. The other Pulldogs look at each other, not knowing what to do. Alan wants to keep a low profile, Dikran and Hakim keep calm, Kenshij is wary: in their experience encounters with the forces of law and order frequently end up as a danger to their safety.

To protect himself from the bright light Nicolas raises his arm. The two officers widen their eyes when they catch a glimpse of what is on his forearm.

"What kind of tattoo is that? Is it moving? And what's that? *Surviving Isn't Enough*. What a load of shit."

"If you don't like it, I'll get rid of it."

Officer Tash grimaces. "Enough games. Now get those animals together, gather your possessions, and leave. Now!"

Alan wants to say something, but Officer Tash is keeping his eye on them like a parent making sure the kids tidy their room properly.

Nicolas makes an open-palmed gesture at Alan to stop. The memory of Little Simon's death is still fresh. No one wants a pleasant evening to degenerate into a tragedy because of a trio of local cops.

The Pulldogs take down their tents and load the sleepy children on litters.

One of the officers – the jumpiest of the three – takes out his pistol the moment twenty minutes have passed and starts waving it around trying to impress them.

"C'mon, hurry up! Tourists will be coming here tomorrow morning and they won't want to have crazies like you under their feet. This isn't a no-global campsite."

After the Pulldogs start their trek along the path leading to the main road, the three officers check over the area with their flashlights to make sure they have not left any nasty surprises behind. At a certain point, while Officer Tash, visibly irritated, removes a large animal poo, a light appears in the forest. At first it looks like a ball of light, then they can see it is a nutshell wrapped in flames, and in the end it lands in the clearing where the camp was.

Officer Tash waves the other two to hurry before a fire breaks out.

The nut is covered in a material that doesn't burn very well. After stamping on it five or six times the fire goes out and Officer Tash can read the words on a flap of skin as large as the inside of a forearm: *Surviving Isn't Enough*.

CHAPTER THIRTY-ONE

Mount Adone

They have been walking without stopping since they left Fosso Bianco.

At midday they are in the Vallombrosa reserve.

When they reach the banks of the Arno, beneath the pillars holding up the A1 motorway, Alan whistles to a fisherman still anchored in the middle of the river, and negotiates being taken across in exchange for fruit. After a little while the fisherman returns with a small raft and ferries them all to the other side.

They follow the course of the winding river, its bends and inlets, for a few kilometers. Alongside the Arno there are a few meters of dirt track covered in rushes, and a footpath passing farmhouses, isolated barns, and country people's sheds. When they meet another walker or a cyclist, the Pulldogs, as if they were normal hikers, ask directions which they then ignore and go straight to where they were told *not to go*.

The act is part of their strategy.

Last week, when they were near Bomarzo, Nicolas had recklessly offered nanites to the farmer allowing them to sleep in his barn. The man had been moaning about the ever-waning harvests, the constantly declining prices of agricultural produce and how nutraceuticals were taking his work from him, but when Nicolas hinted it was possible to refuse the global market's blackmailing grasp once and for all and liberate himself from traditional food, the farmer had given him a strange look.

"This land has been in my family for generations. My ancestors made this place what it is, and in exchange the land made us what we are."

As if this were not enough, Nicolas continued, "I have a great respect for your connection to the land, but in China they are already composing good quality rice in three varieties in nanomats, in one move erasing the ancient agricultural system of the rice paddies. Five hundred million farm workers have hung their hats on a hook and turned to other activities."

"Ah, I saw something about that on the TV, but what do you expect me to do with this land? Yet another B&B or boutique farmhouse hotel like all the others popping up all over the place? Go and ask them if they will let you sleep for free in one of their barns for the night." Then he got up, half offended, and left.

The day after, as they were leaving, he said his goodbyes unenthusiastically.

Once they reach the Vallombrosa reserve the Pulldogs start singing; the music is a tool, like a stick: it stimulates speed and reinforces determination. Anyway, Depeche Mode's 'Enjoy the Silence' encourages them to carry on walking.

Alan is convinced, with the right soundtrack, they will reach Monte Adone the following evening and be able to leave the animals with people who will know how to look after them properly. Afterwards they will have to take stock of these first weeks and decide where to head with their next strides.

They are all repeating the last verse when a shot rips through the air. Then another, closer shot scares everyone. Silvia grabs Alan by the arm.

"Quick, on the ground or behind a tree!"

He raises his eyes just in time to see a shadow disappear quickly into the leaves two hundred meters ahead of them. The first to move is Hakim.

"Wait!" Alan tries to stop him, but he has already gone. "No one move, got it?"

The dingoes growl. Nicolas and Rafabel take Tasia and Pino by the hand; the others hide amongst the trees.

Two minutes go by, then Hakim appears again in the bushes. He is cradling a wounded bird of prey. "Poachers."

"Are they looking for it?"

"Maybe. It's a red kite, an endangered species."

"We have to take it with us then."

Hakim is perplexed. The kite tries to open its beak to let out a cry. He holds it closed to avoid being discovered.

"It's only a flesh wound but I don't know how long it can hold on."

Silvia asks Nico to bring the animal closer.

"Could nanites help him?"

"I don't think so. They can't perform instant miracles. It would take them too long."

Alan agrees they shouldn't attempt fantastical solutions.

"Hakim, do you know how to treat him?"

"I can try, but first we have to find shelter. Those poachers will be on our trail."

Dikran calls the Recovery Center on his mobile, having read on their site that they have a mobile emergency unit.

"Hello? Yes, it's urgent. We've found a wounded red kite. We are in the Vallombrosa reserve."

After a moment he shakes his head. "No, on foot.... Ah, I see."

He closes the call and explains that the emergency unit is dealing with a deer hit by a car on the main road. The only option is to take the kite there themselves.

Gathering the others together, Alan tells them what to do.

"We should divide into two groups. Nico, Hakim, Rafabel, and Kenshij are the fastest runners, they can take the kite to Mount Adone. We will catch up with you as soon as we can, maybe even by tomorrow evening. Even if we bump into the poachers, they won't suspect us of anything."

Rafabel hesitates for a moment, then hugs Tasia and Pino.

"I have to leave you kids. Be good."

The next day Alan and the Pulldogs are welcomed at the entrance to the Wildlife Recovery Center by a woman who introduces herself as Elisa Berto, daughter of the founder.

"What a wonderful place. You have a lot of land."

"Thanks to my parents. They chose this oasis on the slopes of Mount Adone. My father often said, 'Like the earth needs fertilizer to be fertile, animals, including man, need wide open spaces for physical and spiritual well-being.'"

"Wise man, your father. How do you manage everything on your own?"

"We aren't alone. Every year we welcome around forty 'non-operatives', who use A.U.I. to ensure they have a minimum salary.[1] The A.U.I. manages a series of jobs and that undergrowth of requests of the *gig economy* in their name, allowing them not to have to worry about the future. For the state they are still technically classifiable as NEETs,[2] but at least here they feel fulfilled."

"It sounds like an excellent initiative, a bit like that Heinlein book, *For us the living....* Did our friends get here all right?"

The land they are on spreads within the Contrafforte Pliocenico reserve, amidst the valleys of the Setta, Savena, and Zeta rivers.

"Yes, late yesterday night. We settled them in one of the empty stalls. Come, I'll take you."

Some of the volunteer non-operatives come out of one of the containers set up as accommodation units and take charge of the Pulldogs' animals. Tasia and Pino whine a little at the idea of not seeing them again, but in the end they let the adults convince them the animals will be safe here, and looked after by people who care about them.

"How is the kite?"

"Your friends brought it here just in time. Hakim managed to stop the bleeding. We are treating it now and it'll be back to flying, as good as new, in about ten days."

"That's good to hear."

"And the poachers?"

"There were two of them. Father and son. They caught up with us in a jeep. They had no proof but were suspicious anyway. They stole a sheep from us, compensation for the kite."

---

1 Application for Universal Income.
2 People who are Not in Education, Employment, or Training.

"That's terrible. Did they take one of the Merino sheep?"

"Yes, but we got their number plates, and Silvia and me followed them home."

"What happened then?"

Silvia proudly shows her new anthracite black pearl necklace and says, "We opened the trunk of their car and took their rifles. Decomposed them in the nanomat and recomposed them like this…. Now they are useless and pretty."

Elisa bursts out laughing, then, thinking about how stubborn some of the mountain people can be, frowns. "Let's hope they didn't take it too badly."

At first sight the Recovery Center looks like a farm, even if at the start, as Elisa tells them as they walk, it was a hospital where people brought injured wildlife from the area: deer hit by cars, poisoned wild boar, foxes injured by hunters, hares bitten by hunting dogs, sickly birds of prey, mountain goats mutilated by agricultural machinery, and badgers and hedgehogs taken from the woods out of ignorance and then 'set free' without thought. These first guests were then joined by illegally imported lion cubs, maltreated circus tigers, parrots confiscated by the police, and runaway tortoises. The latest arrivals are a tetraplegic poodle, a wolf with a broken pelvis, and a Marsican brown bear wearing a nappy.

The four Pulldogs have settled down into a welcoming box near the wolf enclosure. Only Nicolas is still inside, whittling at a piece of wood.

"Welcome! You took your time. Come in, we'll be crowded but need for nothing. Elisa has been very generous."

"Thank you, Nico. I'll leave you now. I have to go back to doing my thing."

The woman says goodbye and the others settle into the three rooms making up the structure. Alan frees himself of his backpack and asks Nicolas where Rafabel, Hakim, and Kenshij are.

"They are playing outside with the wolf cubs." He looks out the window and points to a fenced area outside. "There is a recovery area all for them."

They all go out. Tasia and Pino are impatient to see the wolves and hug Rafabel again. Silvia, though, hangs back to look at herself in the mirror for the first time in weeks.

Looking out of the window, Alan catches a glimpse of Kenshij stroking a wolf cub, then he turns looking expectantly at Silvia just as Nicolas offers her the piece of wood he has been whittling; she won't take it and goes to Alan in the doorway.

"What's that?"

Nicolas holds out the object.

"I read this sentence one day, I remembered it and wanted to share it."

Alan takes the paddle and reads: *Follow the Sun So You Never See It Go Down.*

"The perfect sentiment for a walker.... I didn't know you wanted to go west. Can I keep it or can you make me a new one just like it?"

Nicolas takes it back and sits down again. Alan thinks this man hasn't learned to control his emotions yet. He sees him take the bandolier from around his waist and put it on the table in front of him.

"I didn't really make it for anyone in particular. It was for me."

"So now you want to chase the sun, Nico? Didn't we agree destinations are illusions and what counts is the journey?"

Nico doesn't answer; he shrugs, lowers his head and turns on his portable nanomat.

"Now what're you doing?"

Alan looks more closely at the display and sees a series of molecule samples in the Nanocad program.

"I'm looking for a cure for anosmia. My father must have used a molecule I'm allergic to but don't know about. I haven't found anything yet. Another possibility is he deactivated the gene producing IFT88 proteins, the ones responsible for the olfactory cilia in our nasal ducts.... No joy there either, unfortunately."

"Good luck, we need your nose." Alan winks at Silvia standing next to him. "Even she thinks your infusions have gone downhill recently."

Nicolas grunts without looking up. "Whatever. They are organizing a party at the Center this evening. Forty-two years since it opened. They asked us if we want to go."

Alan grins. It's been since the beginning of December, since the 'Rishow' show celebrating their defense of the viaduct against the developers, that they've had a party. There might not be a lot of people in the Apennines but it's an opportunity to raise the group's morale after being thrown out of their Fosso Bianco camp and the red kite incident.

"We could play! The twins can't wait to loosen up their fingers."

Nico opens another Nanocad project and lifts the display.

"Great, we can ask Elisa. If she says yes, this is what I want to play."

A mass of different-colored atoms dances around on the screen.

"You play molecules?"

"They're smells really.... The ones I have collected by eye over the last few weeks, using my phone to identify them. I can link them to music and create an amazing effect. What do you think?"

As soon as Alan rests a hand on Nico's shoulder to show his support, Silvia leaves to see the wolf cubs without saying a word.

The resin on the tree trunks marking the edges of the Recovery Center shines like nocturnal road markings. Some of the volunteers, using boards and pallets, have assembled an improvised stage for the concert.

Alan has donned the shoes of his musical alter ego: he is wearing two thick necklaces – one is leather and one is silver – and a spandex suit with the fastening open to his sternum, showing off his well-muscled chest. This time his inseparable guitortoise is ready to loose notes, not arrows. According to what the Pulldogs say about him, Alan is not a great singer; his voice is passable, but his vocal cords release so much energy he is impossible to ignore. The timbre of his voice is powerful, sometimes brutal, and he unleashes it as he jumps, kicks, roles on the ground, waves his arms and makes gravity-defying leaps. His performances meld song, music, and dance more than when he was playing Urban Golf.

One day Silvia paid him the best compliment ever. It was March 31 and he, as a birthday gift, had taken her to run on the beach at the Torvaianica Gates. It was cool and after three kilometers on the sand, he opened a bottle of *Morellino di Scansano* he had been carrying in his backpack, and dedicated an acoustic version of Radiohead's 'Creep' to her. Before making love behind a dune she had whispered in his ear, "Your voice wounds, the perfect voice to make people chase after you."

Now, when Alan sees Silvia circulating with a glass of wine in her hand, wearing a transparent black dress, her mohawk standing up on end, he gets the feeling she wants to shake things up a bit and confuse him. Perhaps she is vacillating between the desire to relive their relationship, transform it into something more stable, and the fear of breaking it up altogether.

This adventure without apparent goal is writing a new chapter in their relationship.

To start the music, the twins begin a tribal rhythm on their percussion instruments, a beat that gets ever more insistent. Alan climbs up on two fruit crates and introduces himself to about sixty people who shout and clap for him. Pietro, Martina's son, is on bass, the display on his instrument shows it has been charged to ninety percent by the sun, ready to let rip burning notes. Behind them, armed with his bandolier full of cartridges, portable nanomat, and a floor fan from Elisa's office, Nicolas is preparing for his first olfactory concert. Alan has no idea what he is going to do; the man is becoming ever more unpredictable and since his creativity has been freed from his father's control, he has been having fun experimenting like a perfect apprentice nanosmith.

Picking some notes, Alan offers himself to the audience.

"Thank you, everyone, I am a little rusty.... It's been a long time since I sang, but this feels like the right time to start again, because in the name of the Pulldogs, I would like to thank Elisa and all of you for looking after us and our animals here at the Wildlife Recovery Center. To make a long story short.... The first song is one of mine, written after a game of Urban Golf."

Alan opens his arms as if he wants to gather the whole Center around him.

Walk and stretch your legs
Listen to me because I won't push you

He doesn't need a microphone: his low, hoarse voice is powerful enough to be heard. A few seconds later, as well as the vibrations of notes and words, the air begins to carry strange scents, dynamic perfumes that change according to the melody and rhythm.

If you can't find the road alone
There is no seat that can save you

Alan crosses the whole length and breadth of the improvised stage. He seeks the eyes of everyone, and everyone is bewitched, no one moves, the infatuation is tangible, their open sensory channels are receiving a synesthetic experience. Some people feel lost, eyes closed, in a reality recreated by the auditory and olfactory melody.

A step forward is worth more than a wheel
A step forward is worth more than a wheel

The ballad receives a round of warm applause and enthusiastic shouting, then without stopping they move into a cover of 'Guaranteed' by Eddie Vedder with Alan's guitortoise filling the stage and Nicolas's nanomat emanating the essences of the places they have passed through, collected during their walk to Mount Adone.

Listening and smelling, no one can keep their heads still, nodding up and down and from side to side; everyone is swaying, hips, legs, inebriated by a narcotic substance taken through the ears, eyes, and nose. Maybe because of the newness and strangeness of the situation, the crowd's curiosity moves from Alan's notes to Nicolas's scents. No one here has ever been to a scent concert, and the novelty doesn't go *unsmelled*.

Head down, long hair covering his face, Nicolas operates the OHMMP,[3] fingers dancing as if in a trance: he moves cartridges and mixes solutions while the display splits valencies, he aggregates molecules and composes

---

3   Olfactory Harmony Molecular Manipulation Program.

smartfumes by eye, olfactory records, and theoretical knowledge, because his nose no longer works.

As soon as Alan realizes everyone is following Nicolas's enigmatic maneuvers he feels a little cast aside and his singing becomes angrier, his yelling even more formidable.

"...and thank goodness it's winter and he's covered up," says Silvia ironically. "If they could see the heliotrons dancing across Nicolas's skin they would go nuts."

"You don't have to tell me," says Rafabel, winking naughtily. Then she pulls down the shoulder of her top and shows Silvia the combined effect the music and scents are having on her skin. The nanites are swarming excitedly, like dynamic goose bumps. Even the patterns on her clothes are changing continually, and at the moment they are covered with an arabesque of Gothic trails.

Rafabel is sitting on the ground in the lotus position, whereas Silvia and Valeria are curled up on a pair of wooden chairs. Glasses of Lambrusco in hand, they are gossiping about the protagonists of the concert.

Even though at the end of the concert he had jumped into a triumph of raised arms, Alan is irritated about how the concert went. Now, pretending to need rest, he is leaning against the trunk of a tree and listening to their conversation: partly because he wants to know what effect his performance had, and partly because he wants to know what Silvia thinks. Unseen, he lights a cigarette, and carries on listening.

"He could have done without the David Lee Roth spandex." This is Valeria's voice.

"I liked it. Now he's back in control of his muscles, it looks good on him. Do you ever think about what the nanites are doing to our bodies?" This is Silvia.

"Hangovers go away more quickly. I don't know if I'll ever get used to slow inebriation," Valeria says.

Alan leans forward and sees Valeria, who has just spoken, finish her glass of wine with a flick of her wrist. Then she starts sniffing an essence and a cloud of lacy vapor leaves her nose.

Rafabel reaches out and asks for a refill of wine. "I don't want to think about it. Before the nanites, you didn't stop to think about what your genes were doing...."

"It's not the same thing." This was Silvia again.

"Yes, it is," Rafabel answers. "Just because you are born with them doesn't mean one thing is good and the other isn't. You've been strange recently."

"Perhaps it's the nanites."

"What d'you mean? Delayed effect? We've been living with them for years; apart from the fever, we haven't turned into monsters."

"I'm not saying they're bad. Only that they change you and you have to adapt."

Another peek and Alan can see Silvia touching her top lip. Removing her anti-smog mask was the first thing she did when they left Rome. She hasn't worn it for weeks. She didn't wear it as protection from the pollution, but to hide the cut warping her smile. No one has ever made a joke about her lip; she was the one who didn't have the courage to show the imperfection in public.

"You see?" Rafabel goes on. "You wouldn't have said such an obvious thing before. I don't think it's the nanites that are worrying you."

"But I do, because they turn you into a different person."

"Is that what is scaring you?"

"You know me, Rafabel. There was nothing tying me to Rome, not even my mother, but I never have had much faith in the future and this means I am attached to the past. I don't know...it's as if I can't let go of the idea of what I was, because I liked *that Silvia*." She pauses for a sip of wine. "Then I watch you guys having fun, playing, singing.... Part of me envies you, part of me doesn't."

"What do you mean? You can talk to us," Valeria says, exhaling little clouds in the shape of infinity eights which decay until they are just strips of smoke.

"I mean, Alan plays, the twins perform too, Nico creates perfumes, and the whole concert.... Great, I'm not saying it isn't, but to me it's like child's play, for people who don't want to grow up."

As Tasia runs past chasing Pino, Rafabel grabs her by the neck of her T-shirt, stops her, and cleans some strawberry jam off her mouth.

"I'm going to say it again, you're being strange. You're the one who is scared, Silvia. You know what you seem like? Someone who doesn't throw anything away because she is sure that she will need it as soon as she does. I think nostalgia is really fear of evolving, of reinventing ourselves, of adapting."

"And sometimes you remind me of Nicolas."

"Huh, you've known him since you were small. He really is someone who has transformed! I don't want to tell you how you should behave, but…if there is any doubt, I'd go for the attempt to throw the old identity away, and I'd get used to it."

"Who knows…maybe I just haven't accepted what I am becoming."

"What do you mean?"

Alan leans out a little more so he doesn't miss a word. In that moment a wisp of smoke slides up Silvia's nose and she turns towards him.

"Gotcha! You're spying on us!"

He comes out into the open with his hands up. "Me? No. I was only gathering fan comments so I can make the Pulldogs' shows even better in the future."

"So it is true, you want to be a singer."

He is managing to keep his tone light, but Alan is worried he has overestimated his own importance, not just as head of the Pulldogs, but also as Silvia's partner. She has never said things like this during their moments of intimacy.

"Am I wrong in thinking I'm not the only person looking for a new identity?"

He leans in to kiss her neck and suddenly realizes he had smelt the exact same perfume on stage a few minutes ago. The association of ideas makes him think of another, terrible possibility. He should have thought about this before. Perhaps Silvia hadn't been listening to his singing during the concert but sniffing Nicolas's smartfumes. Those molecules were kind of sweet and bitter at the same time. Was Nicolas capable of reproducing the smell of a human being on the nanomat? What if he could and that person was Silvia?

"D'you know what I think? The animals are going to do fine here, and we have already seen enough mountains over the past few weeks. I think we should go to the seaside."

Silvia stands, grabs her bag and kisses Alan back.

"That's not a bad idea."

Alan and Silvia say good night to everyone and disappear hand in hand into the dark of the forest to give their bodies some relief.

CHAPTER THIRTY-TWO

In/Semi/Nation

Alan turns his head on the carpet of moss he has been using as a pillow, and moving to put his arm around Silvia finds himself hugging a backpack instead. He hasn't slept much, but it isn't the bed of leaves or January's four degrees Celsius making him feel cold; it is her absence putting him in a bad mood.

Picking up Silvia's backpack, he notices a dark patch on the ground. He touches it and sniffs the substance on his fingers: mercury.

He leaves the forest and returns to the Recovery Center. There is nobody about. It is six forty a.m. and everybody is still asleep after the excesses of the previous night's partying.

In the hospitality stall, the Pulldogs are all sleeping one next to the other.

Alan steps over Tommaso and Pilar sleeping in each other's arms, Dikran is snoring and Rafabel is sandwiched between Tasia and Pino. Nicolas is at Rafabel's feet, lying like a mummified pharaoh.

Hakim is using Alan's backpack as a pillow; with a little muttering he rolls over and uses Silvia's, now right there next to him, instead. Without waking him Alan takes his backpack and leaves the box. Kenshij and Miriam, the only two already up, are doing their morning tai chi exercises in the wolf cub enclosure. They nod hello at him as he passes.

He is going a little further on, to the tallest tree in the Recovery Center. It is over twenty-two meters high and Alan climbs up the trunk and settles onto a branch from where he can see the whole of the valley enclosed by the rocky cliffs of Badolo and the clay slopes of Mount dei Frati.

The wildlife and scenery look gentle and benevolent, as if the valley can absorb some of the anger inside him. He has always liked sitting in high-up places in the first light of morning when the sky is a grayish blue and all the noises and whisperings of the night have gone to sleep. Above all, he likes sitting up high because it doesn't lead anywhere, he can't even climb any higher, from here the only way is down.

He takes three tube-like things out of his backpack and slots them together to make a Tibetan trumpet about a meter and a half long. When he brings it to his lips the long notes of the instrument, oppressive and sad, wake up humans and animals alike.

"Are you sure you don't want to stay a few more days? We can always use an extra set of hands here at the Center."

With the smell of coffee in their nostrils, it is difficult to refuse Elisa's offer. Alan looks at each of his companions one by one, probing their psychological states.

"What d'you say? Does anyone want to stay? We are going to leave the Apennines for the sea, you could catch up with us there."

He has his doubts about Tommaso and Pilar, who are so wrapped up in each other they are already somewhere else both in mind and body, and Martina, the ex-political militant who seems to suffer from some kind of chronic sloth; if it hadn't been for her son Pietro, she would get left behind at least twice a day. The young man, on the other hand, is one of the most enthusiastic walkers and every time Martina lags behind he is there to encourage her and chivvy her along.

The children are low because leaving their animal friends behind is making them very sad, but although they have tears in their eyes, Tasia and Pino shake their heads.

As they are all saying their last goodbyes, Elisa offers Miriam what looks like dark brown focaccia in a napkin.

"It's *castagnaccio*, we make it with sweet chestnut flour from the local trees, and raisins from a farmer friend of ours. It will keep your energy levels up during your trip."

With this the Pulldogs start walking again.

The Pliocene spur is like an open-air geological museum. Fossils and shells prove that between two and five million years ago this was a shallow sea in a tropical climate, with wide bays where the rivers flowing down from the young mountain chains emptied into what are now the Padana plains.

Nicolas has collected so many samples he has had to print a new bandolier to hang across his chest like a gunfighter. He hurries to reach the head of the group and Alan finds Nicolas falling into step alongside him, bare chested and with his hair in two stupid plaits.

"Why the sea, Alan? Weren't we going to decide together?"

"Yes, but we need to take a break. The concert was a success. I loved the way you interpreted the songs with perfumes. We should do it again."

"Yes, but I think we should be heading south now. There's nothing to slow us down anymore. What're we going to do by the sea in January anyway?"

Alan doesn't seem to be in a talkative mood.

"Listen, the red kite convinced me," Nicolas says, not letting up, "in fact it gave me the idea I was looking for. Giving nanites to everyone is no good because lots of people won't want them.... You're right, we should take them to the endangered tribes, help them free themselves from depending on land that is under threat from the global nutraceutical industry."

"I don't know whether to laugh or be worried. First you want to chase the sun, now you want to be like a missionary taking nanites to the wilderness instead of the Bible? Divine salvation printable in 3D?"

"I'm not talking about divine salvation, but conservation of cultural biodiversity."

Alan turns to look at him seriously.

"If you give them nanites you will be making them evolve in a certain direction. OK, you're right, maybe you don't want to be a missionary, you'll feel just fine playing God."

Nicolas walks away to pick up a fossil. The wind and rain have sculpted the rock face so much that the surface looks like a bas-relief illustrating the sedimentary layers. When he falls back in beside Alan, he picks up their conversation.

"God has nothing to do with anything…. But we need a goal. Otherwise, what are we going to be doing over the next months and years? You said the same thing before we left. Where are we going and how are we going to live? I'm just looking for an objective."

Hakim hurries to catch Nicolas up and starts walking on Alan's left. He has to be nearby for any discussion about the future of the Pulldogs.

"You keep thinking we're something we're not, Nicolas. We were anarchists, we had temporary jobs, we were homeless people and punks with dogs; now we are rejects, walkers, perhaps even situationists and bohemian neodadaists."

"You asked me once who I would like to walk with. Do you remember the day we did the Urban Golf trial?"

"Of course I do, I ask everyone who wants to join us."

"Well, I had no idea then. Walking was what caused a horrendous sensation I felt between two periods of comfort. That's why I answered with the first shit that came into my head. Now, though, I would tell you I would like to walk with Thoreau."

"Excellent choice. But you shouldn't think you're the only one with a vision. All the ruins we have been through over the past few days and weeks, the wonderful places we have seen, well, we should experience them more often. We should migrate like birds, from season to season, and stop in places we like. Repopulate abandoned places, without having to live there always, give them back a dignified existence, and share the paths with other walkers. Not being anyone. Each becoming an anonymous singularity. Without managing any property, or any kind of lasting goods or buildings, except for our desires. Without anyone asking for money, without anyone selling anything. Sharing our know-how with other people, sharing the 3D formulas, spreading knowledge, giving back experience. This poverty will make us really rich."

Nicolas says nothing, but Alan knows he is just recharging his tongue. He is proved right because with a spurt Nico spins and starts walking backwards so he can be face to face with Alan.

"You will need a lot of people to make a dream like that happen. People I could bring."

"There are enough of us. Anyway, what makes you think any tribe would accept your nanotechnology as a panacea for all their ills?"

"They have no choice if they want to avoid extinction!"

"Like the Native Americans and the Australian Aborigines, you mean? It's up to them to choose what their future is going to be. And anyway, what do you know about it? Perhaps they will choose to adapt instead of disappearing, like all human beings do in the end."

"If we don't ask them, we won't know. Would you do that?"

Nicolas catches Miriam's attention with a loud whistle even though she is forty meters away.

"Yeah, Miriam, you were there when it all happened," he says, raising his voice so everyone can hear him. "How was Alan adapting to the idea of living confined to a wheelchair?"

Alan speeds up and, without warning, head-butts Nicolas in the chest, making him trip and fall backwards.

"There, now how does that feel?"

The pair roll down a slope and end up in a field of feathery flax, a rare grass which appears to have found its ideal home in the aridness of the rocky spur; if it had been spring they would have seen it in flower, the points of its ears topped with long, feathery tips. If it had been spring, the spectacle would have distracted everyone from the fight, including Alan and Nicolas. It is winter though, and they vanish into the tall grass; for a few seconds there is no trace of them except for the sounds of punching and kicking.

The Pulldogs run down the slope after them. When they reach and surround the brawlers, Rafabel pulls Nicolas away and Silvia holds on to Alan. Miriam looks at them both with disgust.

"What on earth are you thinking? Fighting in front of kids? You idiots!"

At the bottom of the field where there is a dirt track, they can see some people getting out of a car and walking towards them. Two of them are the hunters from Vallombrosa, the others are Carabinieri.

Dad Hunter is agitated and gesticulating at the Carabiniere with a big jaw to hurry up.

"It's them! They stole our rifles."

Alan and Nicolas are still growling at each other and breathing heavily; they haven't noticed the bad news yet. Silvia is the quickest to react. "What rifles? Can you see any rifles around here?"

"The here present Mr. Davoli and son," starts Carabiniere Big-Jaw, "claim you opened the trunk of their vehicle and removed a pair of hunting rifles."

Daddy Davoli hurries to make a point. "But they weren't just ordinary hunting rifles, they were satin-finish high-precision weapons, XLR-12 by Fabarm. The formula and their composition cost me a fortune."

Silvia opens her arms wide and shrugs. "We don't know anything about anything."

Davoli Junior changes the accusation. "So arrest them for vagrancy. We've been following them for two days. First they were sleeping in the forest, pissing and shitting everywhere like gypsies, no, I mean, they...they..." his face creases into a frown of disgust, "...they do a kind of pisspoo, a mixture of shit and urine like guano...like birds, but it's the same thing in the end. Then they hid in the Mount Adone Center with a bunch of wild animals as smelly as they are."

Martina motions to the others to wait a moment, pulls out her phone, presses some buttons and shows a page to Big-Jaw.

"That crime hasn't existed since 1998. Here, look. You two on the other hand killed Lana."

Carabiniere Two – big nose, no lips – turns to Davoli and asks, "Who is Lana?"

Martina continues her attack. "Lana was our Merino sheep, the one they stole from us in retaliation."

"Is that true?" Carabiniere Two needs confirmation.

Father and son say nothing, then Davoli Junior attempts an excuse.

"It was wandering through the woods. We didn't know it was theirs, nor that it was a Merino sheep. Normally we hunt wild boar."

Hakim doesn't let the opportunity go. "And protected red kites like the wounded one I took to the Mount Adone Recovery Center."

Big-Jaw looks at Daddy Davoli darkly and takes back control of the situation.

"Just a moment…. The red kite doesn't matter right now. I might not be able to arrest you for vagrancy, but you are still liable for a fine."

"Why?"

"For urinating and defecating in a public place."

"We were in a forest."

"So? It was a nature reserve."

"We don't have any money."

"So show me your IDs."

Silence. Everybody looks at everybody else. Then Rafabel goes to each of the Pulldogs, collecting a number of identity cards which she hands to the Carabiniere. After flicking through them he turns them over and over in his hand, a resigned look on his face.

"They're years out of date. You do know I can hold you, to verify things?"

Alan tries to defend a somewhat precarious civic situation.

"Think of us as foreign tourists, not resident in Italy, we could be citizens of anywhere."

"Stop talking rubbish. Tourists travel by plane, train, or car, maybe even by bike, and even if they do travel on foot they have valid ID!"

With her usual gentleness, Rafabel lays out her idea. "So take us for stateless people."

Carabiniere Two frowns, not getting it. "What?"

Big-Jaw turns to all of them.

"Stop now, all of you, I've heard enough. Sheep stealing versus rifle stealing. The best thing to do is to forget about all of this, kite included, but above all," he points at the Pulldogs, "it's best you lot vanish from our neck of the woods as quickly as possible."

The Davolis shake their heads but resign themselves to accepting the compromise, however unwillingly.

Carabiniere Two hands back their out-of-date IDs and Big-Jaw takes another look at the whole group. "Do you really want to cross the country on foot?"

The question is prompted by honest curiosity, Alan can see it in his face. A fundamental right like traveling on foot seems to have become uncommon to the point of being easily contestable.

Alan doesn't tell him the Pulldogs have already beaten down biological limits, gone beyond social conditioning and broken cultural taboo. There would also be no point telling Big-Jaw they were planning on crossing geographic and political frontiers. For Alan frontiers only exist for people who don't cross them.

The hunters take their leave with contemptuous words while Carabiniere Two ad-libs as a local guide and gives them directions they don't need.

"I advise you to take an authorized tourist track. The Sentiero degli Dei[4] is not far from here; it's under the jurisdiction of the Centro Alpino Italiano[5] and runs from Florence to Bologna. It is the safest route and has the best views."

Towards evening, before reaching Bologna's Park of Gessi with its chalk formations and looking for a place to camp amidst the Calanchi

---

4   Path of the Gods.
5   Italian Alpine Center.

dell'Abbadessa with its blade-like rock formations, Silvia falls to the back of the line of walkers. Something that Alan can't not notice. Her stride isn't the same as when they started, and for a few days it seems she has been trying hard to think of nothing, as if she wants the inertia of walking to take her over.

In the middle of the foggy dusk there are fireflies flitting here and there, and milky mist is forming in the dips and gorges. They are crossing an area where the hills are riddled with faults, the rock is crumbly and vulnerable to erosion. Alan can remember some rudimentary geology from high school, but it is his experience that has taught him that people, like rocks, are more vulnerable to shaping where they are most fragile, becoming the result of forces that are stronger than they are.

The small Emilian canyon has formed where the rock is less compact, and the towers of rock – over five meters tall – have become isolated from the cliffs for the same reason, selective erosion. Wind and rain have created less erosion where the layers of rock are more tenacious at the top and along the sides.

Alan would like to be able to tell Silvia the well-being of each individual is at the root of that of the group, and after their night together in the wood he would like to find a way to make her understand he cares about their relationship, that his decision to go to the sea is for her, her insecurity and fragility. On the other hand, he doesn't want the responsibility of that decision to weigh on her. Even though they are not family in the traditional sense of the word, the bond connecting them has been forged through dramatic experiences like the incident at Globalzon, and the months of rehabilitation and physiotherapy, Little Simon's death and the closure of Il Romoletto, but also through exciting experiences like the games of Urban Golf, starting the colony on the Garbatella-Testaccio viaduct, and defending it from the speculators.

With a sign he calls his mother, Miriam, to him. Without going into details, Alan explains the situation and asks her to find out what is worrying Silvia.

"I'm going to call your mobile, you keep it on in your hand. She won't notice anything."

Her phone rings but Miriam isn't too happy about playing spy.

The rolling hills, meadows, and clay ravines have given way to the woodland flora of the Apennines. Miriam doesn't have to slow down much to find Silvia at her side.

"How are you, sweetie? I've never seen you back here with the stragglers before. You're always at the front exploring the way forward."

"I'm not feeling so well."

"Is it a physical thing?"

"Partly, but mostly just a feeling of overall discomfort. Have you seen how those two keep arguing? Nicolas wants to start the nanite revolution and save the world, Alan wants to go on tour with his band in mystical locations. But what do I want? I worry that the more we go on, the harder it will be to take everything back if we find we are on the wrong path."

Miriam looks at her as if she knows what is worrying her. "This confusion will pass soon...you'll see."

"I'd just like to know when."

"A little rest and the sea will do you good."

"Yeah, I'm so pleased Alan remembered what I wanted."

"He did it for you. Because he cares about you."

Miriam shoves her mobile into her trouser pocket and holding her finger over the microphone says, "Y'know, the others might not see what's happening to you, but I do."

Silvia links arms with Miriam and slows right down, almost stopping, to leave some space between them and the rest of the Pulldogs.

"Really? How did you know?"

Further up the line Alan shakes his phone, he can't hear them anymore. He peers backwards trying to see what is going on, but all he can see is his mother whispering in Silvia's ear.

"Little things, you eat strawberries, you avoid certain infusions, you talk more than usual and walking tires you. The vomiting isn't because you are expelling toxins. I mean you are, but your nausea is because you are pregnant."

She rests her hand delicately on Silvia's still-flat abdomen.

"How many weeks? Nine?"

Silvia nods almost imperceptibly.

"What I don't understand is why you don't want to tell anyone."

"I will, but I haven't found the courage or the right words yet."

Miriam turns and heads back to the front of the group.

"I'll keep your secret. Alan is worried, but I'll only tell him the minimum necessary."

CHAPTER THIRTY-THREE

The Green Ark

From a chalky crest in the Vena del Gesso national park, Alan studies the scenery made up of green-on-green hills standing out against the sharp blue of the sky.

"Down there, that way, as far away as possible from any towns or villages."

The segments aligned on the horizon, gray like layers of dust, are towns: Imola, Castel Bolognese, Faenza, Forlì, and further east, Cesena. Beyond the dividing line of the A14 motorway Alan can imagine other towns and cities: Modena, Ravenna, Ferrara, and Bologna, black holes of energy bristling with light and saturated with noise, smog, and sewage infecting the air with urban flatulence.

As the sun's circle rises above the tops of the trees, the Pulldogs reach an old mill which, given the rotting wood piles and the remains of a farmhouse, looks like it must have been lived in until quite recently. A continuous gush of water from the mill pours into two large cisterns.

The Pulldogs gather around a drinking trough. They wash and fill their camel packs using telescopic straws which suck up the liquid in a few moments. Then, as they sit in a circle nibbling on fresh early produce, Dikran takes the opportunity to show the others their planned stops for the following days.

"To keep under cover of the trees we're better off walking parallel to the A14 until we reach Cesena. From there there's an open stretch through the middle of the Padana plains which we cannot avoid. There are no shortcuts or alternatives, unless one of us has a friend or contact who can help us." Everybody shakes their heads and Alan can only agree.

"As well as being the safest route, I think it is also the shortest."

Rafabel is spreading suncream over Pino's and Tasia's exposed skin. Even being under the pale winter sun reduces their daily autonomy, with the risk of having to stop more often to find water sources.

"So where are we going to go by the sea?" she asks. "When I was a little girl my parents used to take me on holiday to Bellaria, a peaceful place, perfect for children. But that was in 2000, and even then there wasn't a wood to be seen, let alone now. The only green space was inside a building that probably used to be a sanatorium. I think this was near the Uso River. There was a small pine wood, too."

Dikran checks Rafabel's information on his phone over the mesh net.

"You are right, that's the only green space that is still public between Cervia and Cattolica. The Parco di Levante is now a HappyLand, and the Parco di Ponente was bought by the Atlantica company five years ago to expand their already existing water park."

Rafabel is shocked. "Do you mean that along, what, sixty kilometers of coast, there isn't even one patch of woodland?"

"Fifty-five, to be precise, and yes, no woodland."

Alan stands, shrugs into his backpack and picks up his guitortoise. "Good, I mean bad…so our destination is decided, let's just hope there are no more unpleasant surprises."

"Miriam!"

Every time she hears his voice, even over the telephone, it puts her in a good mood.

"Ivan Mihailovich Shumalin! How nice to hear from you. Where are you calling from?"

All the Pulldogs are now gathering around Miriam and she puts the call onto speaker phone so they can all hear.

"Ah, my dear…I have finally come back to the Mediterranean. We are sailing from Cyprus towards the Aegean."

"Really? We are heading towards the sea, too, but the Adriatic."

"I knew that, I've been following your route on iMaps. Your son updates the paths you have covered every day. I know it isn't the same thing, but by following his journey logbook I feel a little like I am there with you."

"But you are coming back to Italy. Can I ask you why?"

"Big issues, Miriam. Hasn't Alan told you? It's a project I have been collaborating on for a few years, the Green Ark. If everything goes for the best, we have launched the first of many vessels."

She darts a look at her son who is miming that he has told her everything he knows.

"Alan has only told me it's a project of mass distribution cooperation."

"Oh yes, an enormous mass," he says jokingly. "You should see it. You would be speechless."

"So, will you stay for a while?"

"That's why I'm calling you. I'm sending you a link to our route. It might take another few weeks, but if you want to leave Europe we could give you a lift."

"And the Naval Blockade?"

"That won't be a problem for us."

"All right then, I'll discuss this with the others and then we can talk again. A big hug, Ivan."

"I can't wait to hug you again, and Alan, and meet the rest of the Pulldogs in person."

A sea breeze coming in gusts blows scraps of greasy waste paper around and makes empty cans roll along the sidewalk. There is no one around in Bellaria.

Rafabel takes off her shoes and hangs them from her belt to let the cool sand stroke the soles of her feet. Then she runs to the edge of the water, Tasia and Pino chasing after her. A moment later all the others get there too.

During the day the winter sun got stronger and the Pulldogs only leave the ex-sanatorium's woodland to bathe in the calm waters of the Adriatic when the brightest stars win their daily battle against the artificial lighting.

They have been doing this for about two weeks, following the rhythms of nocturnal animals. By day they rest, they keep to themselves and try not to stand out, whereas at night – when the seafront is deserted except for joggers and the rare cyclist – they cross Via Pinzon, climb over the fences around the private beach 'Polo Est' which is closed for the winter, and dive into the sea.

"When is the ship going to be here?"

During their swim Tasia repeats the question, as she does every night.

"Soon…it has to come a long way and it can't go fast," Rafabel tells her.

Pino is playing with two toy ships Nicolas taught him how to make on the nanomat.

"How will it get through the Naval Blockade? Will they be shot at?"

Suddenly appearing from deeper down underwater, Alan surprises the children, making them scream happily.

"I hope not. As far as I know the Green Ark is like a tiny nation. It can't be stopped, you can't shoot at a nation just like that, just because you want to. As long as it stays out to sea no one can do anything to them."

Tasia wants to climb on his shoulders so she can jump into the sea.

"Why is it called the Green Ark, Alan? Are there animals on it, like on Noah's Ark?"

"Hmmm, I don't know. As soon as they get here we will find out."

The little girl jumps in and when Alan turns to the shore he sees Miriam talking on the phone.

"Look over there…. Maybe Ivan is calling right now."

"Are you anxious?"

Silvia takes the hand Alan is offering.

"No, after two weeks of resting I feel much better. I'm happy really, I'm glad the appointment is in Rimini, so I can start walking again."

She hasn't vomited for a few days and Alan feels this is partly down to him.

"It's sixteen kilometers to the western docks. We will meet with other people who want to board with us there."

"Has Ivan told you how many?"

"No, and the ship won't dock, it will send dinghies to collect the people."

As they walk the waves begin to get choppier in the breeze. Out to sea they can see the lights of the pitching night fishing boats. Some passersby stop to look at the Pulldogs curiously: a small procession walking along the seaside isn't unusual, but after midnight it still catches the attention of people who have nothing else to do.

Three hours later, after passing the docks and dozens of hibernating vessels, they reach the western quays. An old man is sitting on a bench next to a younger woman. The man is holding a walking stick, and his bearded face, above a raw linen suit, accentuates his austere bearing. He has long, unbound hair and his fingers are tapping out a tattoo on the wood of the bench. His eyes are staring at the horizon, so absent he doesn't even notice the presence of the Pulldogs.

The woman beside him, on the other hand, notices them immediately and smiles. She has wide shoulders like a swimmer and a graceful face with the high cheekbones and elegant features typical of Somalian women. Her high forehead is topped by long hair braided down her back.

"Good evening, or rather, good day. I am Farisa Haabil, and this is Mr. Paolo Salandri. You are the Pulldogs, is that right?"

After some rapid introductions, Farisa helps Paolo to stand up.

"Ivan told us we wouldn't be the only people boarding," she says as she looks at her watch. "The dinghy should already be here. Paolo wanted to come early...travel anxiety."

The woman opens her bag, rummages around for some medicine, opens a pillbox with no label, and slips a tablet into the old man's mouth. He makes no sign of discomposure, swallows, and continues to stare at the horizon.

Then the children begin to shout, "Is that it? Is that it?" They have seen a trembling dot of light on the dark horizon.

Ivan Mihailovich Shumalin throws the end of a rope to Alan, who grabs it and pulls the dinghy to the quayside.

"Ciao! Finally we meet again. C'mon, jump on board."

As they file past him, Ivan hugs each and every one of them. He is especially warm towards Miriam and Paolo Salandri. Then he goes back to the helm of the dinghy and waits for Alan to jump in.

"Hold on tight. It's time to fly."

At a few kilometers from the coast the Green Ark looks like a tangle of luminous lines, an imposing bulk outlined against the dark water.

"Wow! That looks like...like an aircraft carrier!" Despite the dark, Dikran takes a few photos and searches the Internet for information about the vessel. Ivan preempts him and tells the story of the ship.

"Yes, in its previous life it was supposed to become a warship. Actually, it's a decommissioned aircraft carrier we bought and modified through a crowd funding and distributed mass computing project. The vessel used to be the *Variago*, the second biggest Admiral Kutnetzov class aircraft carrier, launched in 1988 from the shipyards in Nikolaev, Russia. In 2004 it was re-baptized *Liaoning* and sold to China, officially as a ruined hulk, so it could be towed through the Dardanelles. Just think, it was supposed to be turned into a floating science fiction theme park, a gigantic fairground like Disneyland, but fortunately this didn't happen and we turned it into a botanic garden to preserve biodiversity, like the seed deposit on the Svalbard islands in Norway, only we don't freeze our seeds, we share them."

Ivan laughs while the pilot points the dinghy towards the mooring point.

"This also means we can move freely, because we are a scientific project approved by the United Nations. Anyway...well, it is a forest, trees planted in earth, even if it is enriched with nanites to improve their health and resistance to parasites. In fact, we are fighting GMOs with nanites: unaware bad genes against good and totally aware NEMS."

There are a number of other smaller vessels connected to the *Liaoning* by gangplanks, making the whole thing a semi-mobile archipelago.

"The Green Ark is a sustainable arcology and after being admitted to the circuit of micronations has got its own flag," Ivan says, pointing up to where a spotlight is shining on a flag depicting a white redwood on a stylized ship against a green background.

The arcology doesn't so much sail as move like a giant raft of organic material. It covers hundreds of meters in every direction and near the keel, below the surface of the water, despite the dark of night, they can see long filaments like leafy branches and thick roots like pipes descending into the deep.

Everyone leans out of the dinghy to see the strange sight better.

"What are those, tentacles?"

"You could say so. Those water roots contain nanobots and collect substances that are useful to the plants and release our biodegradable waste into the ocean. The whole vessel is covered with a layer of vegetable fibers and nanites functioning like sapwood and duramen."

Nicolas looks very interested and asks something that sounds banal at first.

"So the structure grows every year?"

"Exactly, the nanites emulate the behavior of vegetable cells."

"Morphogenesis."

Ivan shoots a look at the man who has just spoken. Miriam had told him there was a walker in the Pulldogs who was ingenious with nanotechnology.

"I see you know what you're talking about. We use morphogenetic models to guide the growth of undifferentiated cells, a little like plant seeds which from birth possess the development plan of the whole plant they will become. We started with the works of the architect John

Johansen. It took four years of experimentation carried out by various groups spread out all over the world."

Curious, Alan slips his arm into the sea and pushes his face close to the surface. To begin with he can only see the wrinkled water and the reflection of the moon. Then he sees a milky smudge which resolves into lots of little lights to become a tail swaying like seaweed. The root brushes Alan's arm and closes in on itself, then it opens again and takes on the shape of a glittering tentacle. In the end it sinks and disappears under the hull of the Ark.

"I've never seen anything like it," says Alan, touching his arm.

From the underwater darkness, the glimmering entity returns to the surface; this time it splits into four or five roots even brighter than last time. They flash about and reform, their light intensifying. The passengers instinctively pull back, worried by the tentacles pushing their way upwards ever faster and faster. After a moment the whole surface erupts in a mass of bubbles.

"Don't be scared," Ivan hurries to say. "It's the Green Ark's way of welcoming you aboard."

Then, as they are about to dock with the larger vessel, a high, strident noise comes from behind them. Everybody looks about blindly for the source of the noise but it is almost impossible to see more than a hundred meters. The shriek comes again another two times, until Hakim realizes what it is and holds out his arm to let the red kite land.

"You escaped from Mount Adone! How did you manage to find us?"

The bird opens its beak and seems to want to scream out its happiness.

CHAPTER THIRTY-FOUR

Mare Monstrum

From the parapet of the Green Ark, Alan watches the foam trails they and the other vessels are leaving behind them: a ferry, probably heading to Igoumenitsa or Patras, two cargo ships carrying containers, and some rusty old barges with flags so faded they are unrecognizable. Further out a cruise ship, loaded with tourists, is moving as slowly as the Ark. In the distance a dark smudge of smoke blights the horizon. Something is on fire, perhaps it is a yacht that has been boarded by pirates who are burning it with purifying flames.

Trails of waste are crisscrossing the waters of the Adriatic. Plastic bags, bottles, polystyrene boxes, scraps of wood, and shiny slicks of black or rainbow oil seem to be following any number of marine routes.

Still half asleep, Alan turns to the deck. His friends, spread through a forest of leafy trees, are all sleeping. They aren't the only ones, there are hundreds of people sleeping there, some are members of the crew, some are strangers in transit. The previous night they hadn't been able to look over the whole ship or floating forest because, tired and weary, they had settled down to sleep immediately, but now, in the morning light, they can look around.

Three artificial hills with a stream running through them recreate three microclimates: temperate, desert, and Mediterranean. The habitats are protected by geodesic domes about sixty meters high; their gradients make the surface transparent or opaque depending on what is required. Beneath the control tower there are two enormous 3D printers, at a guess their length and breadth are twenty meters each, a desalination apparatus, and other devices Alan doesn't recognize.

Attracted by the singing of a goldfinch, his eyes move to the Mediterranean habitat where, amidst the tree trunks, dozens of drones, like little birds, are fluttering here and there watering where needed.

A man comes up to Alan. He has Slavic features, blond hair tied back into a ponytail, and his lips and nose are rather prominent. His ears are pierced and are home to two devices like round batteries.

"Welcome aboard, I am Kirill, one of Ivan's friends. You must be Alan."

"My pleasure, yes that's me. Thank you for your hospitality."

"No problem, consider it an exchange of courtesies, your country was my home for years. I was working with Ivan when we won the Nanoblock prize. I was on the team, but then, when he was arrested for copyright violation, well…I couldn't cope, and I left. I went back to Russia, but now we are working together again, on the Ark."

Nicolas and Ivan come out of the forest behind Kirill; it looks like they are getting to know each other. Behind them Miriam also appears. She is clearly fascinated by the flora surrounding them.

"Kirill, have you met Nicolas?" Ivan asks enthusiastically.

They shake hands vigorously and the Slav notices the pictograms on Nicolas's forearm.

"Not yet, but his skin behaves well in the sun."

Along Nicolas's arms the heliotrons are sliding around, capturing energy and emitting a silvery gray light.

"You should see how he has modified the nanites," Ivan says happily. "It's just what we need to improve the genome of our plants."

The surface of Nico's skin is red and shows signs of scaling and blistering on the parts most exposed to the sun. In general his skin looks dry and shiny, it is hot to the touch, and in some places it is thickening. On the dermis, the hair follicles are sparse and there are filaments interwoven with the connective tissue. It looks like a compact network of elaborate laminar membranes.

"How did you do that?"

Nicolas puts his arm down, makes two tight fists and then shows him the palms of his hands. The heliotrons react in real time, running to his fingertips, so much so that his fingers go dark.

"I think I did what you have done, the trees inspired me. Contrary to the cells of a photovoltaic panel, tree leaves compensate for the deterioration caused by prolonged exposure to the sun by recycling their proteins every hour. Their cellular repair mechanism allows them to use energy without losing efficiency. By imitating this ability, the heliotrons work like auto-assembling molecules capable of freeing themselves of damaged photons in the form of electrons and therefore electricity. In practice these are carbon nanotubes with a structure, a means of electrical conductivity, with synthetic phospholipids to help, enabling my whole skin to collect energy in a uniform way."

"Side effects?"

"For the moment recurring itchiness and a temperature, sometimes quite high. I'm expecting some surprises in a few months."

"Why?"

"Because of these areas of thickened skin."

Nicolas probes his elbow with his fingers where the skin is redder and drier than elsewhere. Translucent oval scales, the size of a coin, alternate with opalescent flakes of skin.

"Do you have to keep it moist?"

"If I could still smell I would be able to find the right materials to compose a moisturizing ointment..." Nicolas rubs his hand over his nose, "...but I haven't been able to since my father left me a terrible legacy, anosmia."

Kirill rests his hand on Nicolas's shoulder, as if offering him condolences.

"He's not really dead, although he is to me."

Ivan immediately tries to lighten the tension.

"Come to the infirmary later and I'll find you something to hydrate your skin. In the meantime, with these heliotrons, in ten years we will all either be extinct or improved. Who knows how many modifications by how many designers have been made to the original nanite prototype. The day will soon come when technology goes beyond our ability to watch over it. That day technology will have come of age and will know how to look after itself, then we won't have to worry about it as if it were about to attack our survival, we will be able to simply keep an eye on it, and keep it in check if and when something bad happens."

Kirill points to the horizon.

"Be optimistic, my friend. Look over there, that is something bad. Every time I see it I get goose bumps."

A solid black line occupies the horizon joining Otranto in Italy and Orikum in Albania. The bulkheads of the Mobile Naval Blockade rise smooth and impregnable, about twenty meters high; easy to put together but difficult to scale, they represent the sticking plaster solution to the problem of immigration into Europe. Similar bulkheads have been placed between Pantelleria, Lampedusa, and Malta, forcing the vessels coming from North Africa and the Middle East to follow longer, more expensive routes. In a few days whole tracts of sea can be closed off and sealed, channeling all boat traffic through navigable checkpoints. The same thing happens on land where the use of surveillance drones high in the mountains, and high blockades on the plains, seal Europe's borders.

Alan puts his hands in his hair. All borders depress him and here is a wall built with the sole purpose of separating international from national waters; it makes him feel even more demoralized.

"I've only seen that on the web. It's monstrous."

Behind them Farisa, wrapped in a tribal-print shawl and accompanied by a limping Paolo Salandri, appears and Ivan hugs them warmly once again. The woman leans against the railing solemnly as if she is about to declare war on the world.

"I've seen two of my friends drown beneath those blockades." No one says a word until Farisa starts talking again. "That's why, when he asked me, I decided to accept Paolo's proposal and liberate him. Even though he is one hundred and two, living what he has left of the present will make it easier for him to forget the past. It's what my father often said to me in Cameroon."

Ivan is moved. He strokes Paolo's wrinkled face, and he is unable to hold back the tears.

Alan has seen him sitting looking out from the prow, watching the horizon, staring at the waves and listening to their movement – breaking and spraying against the hull of the Ark – almost hypnotized by it all. Another time, he has seen him in the stern as if enthralled by the trail the ship is leaving behind it, the traces of their passage.

"Thank you, Farisa." Paolo sits on a bench as he speaks, and rests his chin on the handle of his walking stick. His eyelids look heavy and he has two large bags under his eyes. "I could never have done this on my own."

Ivan sits down next to him.

"Perhaps you don't know, but this man is Prometheus. Without him the Nanoblock competition would never have seen the light of day, and the Ending Hunger project would have remained an idea, not to mention the nanite prototype developed for the composition of nutritional substances…and he offered to be the first guinea pig. We all know how that ended."

"I didn't do anything special. It would have taken more bravery to carry on life as it was before. I was sickened, it's not New Age philosophy. Do I look like someone who needs to find himself? When it comes down to it, the more things you can do without, the richer you are…."

Alan looks dark and at the same time astonished at realizing his life is the result of strangers meeting and unimaginable circumstances. A part of him is happy, another part could cry.

"I simply initiated a process that was already in the air. The Master did the rest and now you are all carrying the consequences of our actions. I hope this path of transformation takes you a long way."

Alan and the rest of the Pulldogs look worried, crowded on the main deck, when they pass the barriers of the Naval Blockade, escorted by a swarm of military drones.

"It's procedure. They are scanning us," says Kirill, "to see if anyone has left the boat. We are all registered...registered so we can be free." He ends with a dollop of sarcasm.

"What about us?"

"You do not exist. And they don't care, whoever came in on the Green Ark has to leave on the Green Ark. They don't care if we are leaving Europe with more people. But, if you try to come back in, well, that would be trouble."

"Like visitors to a prison."

In Alan's eyes the Pulldogs will only be able to start a new community when they are far away from Rome, Italy, and Europe. This long gestation – from the foundation of the viaduct to here – has reached the point of giving birth to a new existence for them. It is too early to say what will happen, but his mind runs through a number of bizarre scenarios. Will they continue to migrate along established routes, like schools of fish and flocks of birds? Or will they drift in the wind like pollen, trusting to find fertile land in every season? In the mists of the future, he thinks he can almost see the outlines of the highest peaks, a lifestyle and behavior that will give shape to a new society, but what lies in the valleys remains mysterious and unfathomable.

"Ciao, Alan." Silvia's voice brings him back to the present. She takes him by the hand and leads him to their tent, where she sits down in the lotus position and strokes her stomach. She smiles and lifts her T-shirt to expose her abdomen.

The morning light filtering through the tree branches plays over her face.

"I have something to tell you," she says, seemingly illuminated inside and out by an intense light. "We are out of danger now."

CHAPTER THIRTY-FIVE

Microships

The storm mass grows until the eastern sky is fractured by a cage of lightning bolts. The cloudy front from inland Peloponnesus is approaching slowly, like a wounded animal, head hanging low it bends forward, thunderous and obscuring every glimmer of sunlight. Bolts of lightning shoot down to the water, ten thousand volts at a time, crackling down a segment of ionized air and flashing at eight hundred kilometers an hour. The wind refreshes the air and rain pours down on the Green Ark, drilling dark marks into the surface of the rough sea.

Many of the people aboard have taken shelter under a tent-like structure to watch the water, light, and air show from a safe area, while others are dancing in the torrential rain getting invigoratingly wet and singing and laughing. The birds aren't flying a salute for the event, they have vanished, hiding in their shelters in the domes.

The peals of thunder come again and again. Every time is like an explosion in their guts, the hairs on their arms rise and the hair on their heads is electrified; all the metal objects start vibrating like tuning forks.

The geodesic domes channel the water into large streams which gush onto the main deck, leaving muddy splodges. Visibility is low but Alan still recognizes Hakim, sitting cross-legged, head leaning to one side, meditating. Nicolas is also out in the rain, motionless, bare chested. The rain is falling on his wet hair and runs down his shoulders and outstretched, slightly raised arms. He looks as if he is welcoming as much rain as possible, filtering it through his open hands.

"Just look at him, he thinks he's Jesus Christ," mutters Alan to Silvia behind him.

She comes to stand next to him at the tent's opening. Laughter escapes her lips when she sees their friend because there really is something Christ-like about his posture, or perhaps the ecstatic expression on his face.

"Perhaps the tribe-saving mission has gone to his head."

Alan and Silvia go back inside and lie down on their mattress, gazes turned up towards the opening where they can make out a portion of the dark sky. Their hands entwine lower down, on her already visibly swollen stomach.

"I can't believe it…I think I can feel it."

"Don't believe it, that was thunder. At seventeen weeks the fetus is such a tiny little being, only a few centimeters, it can't kick."

"Don't call it a fetus. We have to find him a name."

"Or find her a name. I think it's going to be a girl."

"How do you know?"

"Female intuition."

"All right, so a double choice. It's nearly time to dance. Are you ready?"

"Are you talking about my pregnancy or the storm?"

"The pregnancy."

"What's the matter, Alan? You have never been pro-children. It's taken me weeks to get up the nerve to tell you about it. After all those arguments we had about this, I was expecting a different reaction from you."

A sinister light turns the sea white. Waves four or five meters high chase each other until they attack the perimeter of the Ark. A wall of water spray, created and carried by gusts of wind, makes the branches of the floating forest shake and rustle.

"So many things have happened. For example, I've realized my future isn't proceeding at the same pace as the world. Our transformation has pulled my skin, straightened my back, and mixed up my internal organs. But the most important thing is what hasn't happened yet."

"What do you mean?"

"I know it might seem banal, but relationships start to go wrong when you only concentrate on the details and lose sight of the general project. Walking has helped me reflect, and, walking, I've discovered there is a relationship between what we can understand with a look, and what goes through our heads…"

A shock wave forces them both to hold on to the tent's structure.

"…big ideas need big panoramas and new horizons. And now I know, when I listen to music or look at a row of trees, when I watch the clouds passing by or I am distracted by a river, my thoughts are different, they don't get bogged down in reasons, they don't get lost on problems, they don't run away from problems. The viaduct was a fundamental period, but it was static, always the same, unchanging, and now it has gone. In its place there is an empty space to fill."

"Do you mean an emptiness between us?"

The rolling increases. Outside people are shouting joyfully.

"Partly, but I mean more of an identity void. Becoming a father might be what fills that gap."

"Might be?"

"No, sorry. I'm certain."

Silvia is surprised by his words and holds on to him, only partially to protect herself from the juddering of the ship.

"All right then, but maybe we should talk about this in less dramatic circumstances, don't you think?"

The pitching of the boat gets more annoying every time the Ark falls into a deep trough between the waves. The noise is getting louder, a cacophony made up of the wind's fury, the pounding of the sea and vibrations like the rolling of a drum beating out a charge. The links to the vessels being towed are put to the test by the assault of the full force of the elements.

Suddenly a voice shouts over the roar of the weather, "Man overboard! Man overboard!"

Kirill points out someone in a rowing boat coming out of nowhere. Then he orders his men into a recovery maneuver.

"Throw him the ropes! Pull him in! Quickly!"

After a number of unsuccessful attempts, the man manages to grab the end of the rope and is pulled on board moments before his small boat smashes against the hull of the Ark.

The storm has no other surprises for them.

Once safely aboard, the man tells them he has been drifting for days. His name is Shimbo Dalisay, originally from Manila in the Philippines but living in Naples; he had been working on a fishing boat hired by five German tourists when they were attacked by pirates. He managed to escape but the storm would have killed him if the Ark hadn't pulled him to safety.

"Are there many others?"

Kirill offers the man, shaken and worn out by the storm, a cup of tea.

"Yes, they follow the tourist routes. They attack sailing boats, yachts, and small cruise ships. I heard they have a base on the island of Antikythera. From there they can attack whoever passes through the Aegean."

"Are they well equipped?"

"I don't know about that, but they aren't beginners. I've been doing this job for four years and I have crossed paths with them three times. This time it didn't go well. I think they know the commercial routes and the movements of the container ships. There are probably regional Mafia families behind them, mercenaries paid by some company, the underworld in general. Whatever, they must have support inside the maritime companies, the harbor masters' offices, and the traffic control organizations."

Kirill looks at Ivan worriedly as he sips his tea.

"What do you want to do, Shimbo? We could leave you on an island from where you can make your way back home."

"Can I think about it?"

"Of course, there's room here for everyone. Just, well…you won't find any kitchens. You'll have to adapt, like we do."

"How do you mean?"

Ivan points at the geodesic domes with fruit trees.

★   ★   ★

An hour after sunup they can see flashes on the horizon.

"I can see it, Ivan. Are you thinking what I'm thinking?" asks Kirill, at the Ark's helm.

"I'm afraid so." Ivan passes his friend the binoculars.

Next to them Alan and Silvia are holding hands, both look worried. In roughly a month of walking they haven't had to face any really serious problems.

"It's hard to make them out.... From this distance it looks like there are a dozen of them, but there could be more."

Ivan slows them down and turns to starboard to earn a few minutes to organize themselves.

"We can't go around them, nor attempt evasive measures. If they are heading for us and want to board us, we'll have to invent something to keep out of trouble."

Nicolas, Dikran and Hakim arrive on the command bridge.

"What's going on? Why have we slowed down?"

Ivan hands Nicolas the binoculars.

"Pirates?"

He looks for a few seconds then hands the binoculars on to the others.

"Yes, even if it would be more accurate to call them *peerates*," Ivan says. "They take advantage of social networks, geo-referenced maps, and 3D printing to board the boats they come across."

Dikran zooms in with the binoculars. Lots of boats, no bigger than motorboats, are approaching fast. They look composed, following the specs of actual patrol boats from the navy, just the way they look at the nautical boat shows or on specialized websites. Every now and then, after raising sprays of water, these sinuous cobbled-together vessels, somewhere between turquoise and green, disappear underwater, stay out of sight for a few moments and then reemerge, like a school of sharks.

"*Jugaad*," Dikran murmurs. "They've adopted the Indian custom of building something using what they have at hand, and updated it to suit modern times."

Nicolas takes the binoculars back.

"Let me see again, I've heard about them. Excellent designers."

Shaking his head Alan tries to stimulate Nicolas's creativity.

"Apart from your admiration for them, can you think of any way to stop them? A trick with molecules from your OHMMP, or a phial? Like the stink bomb you made during the siege of the viaduct?"

"The area is too big, and they are too fast. It would have no effect."

Nicolas can see them gathering in groups of four or five, then dividing again so as not to become a target, then they submerge and reappear somewhere else: agile and fleet, they act with a shared intent, hunting and pillaging anything that floats.

"In reality we have already come across them once or twice, off the Turkish coast," says Ivan. "But there, with all the sailing boats and yachts around, they had juicier and tastier things to grab."

Silvia rests her hand on her stomach. "So what are we going to do? You're not going to wait for them to board us, are you?"

Kirill rubs his earlobes roughly, a gesture halfway between a nervous tic and a call for good luck. The metal plates in his lobes become iridescent.

"This is a research ship, and it has never been an aircraft carrier."

The man takes the microphone and opens a line of communication with the fleet of vessels following the Ark.

"We have company. Pirates on motorboats. Prepare to defend yourselves." Then he sends a report to the Coast Guard. "Mayday, mayday, mayday, this is ARK, alfa-romeo-kilo. We are being attacked by pirates. Our position is 36°11'19.9" Latitude North, 22°41'07.4" Longitude East."

Tracer bullets rise from the motorboats, drawing their dangerous arcs in the air.

White smoke floats around the tents, dampening the shouts and impeding visibility. The hull of the Ark hasn't been much damaged; there are dents along the surface but no holes. The pirates have fired some shots, but concentrated on the trail of smaller vessels, and the reckless men who tried to board the Ark by climbing the sides of the hull using grappling hooks, harpoons with epoxy resin, and extendable ladders have all been pushed back into the sea.

The Ark has shown itself to be an impenetrable fortress, not least because the defenders have a twenty-to-one numerical advantage.

From the railings Alan picks out the head of the attacking swarm and sees him, frustrated by the failure of the operation, pull on a gas mask. His outline is just visible through the banks of mist wafting here and there. Without thinking it through, Alan loads a dart onto the guitortoise, aims at the head of the swarm, but can't get a good shot. The shadow flickers in and out of sight like a frayed cloud. Then the man bends over and pulls a bazooka out from under his seat and sets it up on its tripod.

Alan feels someone grab his arm. A face emerges from the milky fog, the lower half hidden by an anti-smog mask. It is Silvia.

"Keep down!"

The head of the swarm launches the first shot, to create panic.

She throws herself to the ground, dragging Alan with her. "Close your eyes!"

Her warning came too late. He sees the explosion destroy a portion of the geodesic dome covering the temperate microclimate. A spark gets inside and starts a fire amongst the topmost branches of the trees.

Dozens of people run to put out the flames using the hoses from the watering system. Nicolas and Dikran are each aiming a jet of water at the flames, making steamy sprays they hope will manage to suffocate the fire.

Another mottled wake hisses in arrival from the same direction. Alan sees this one coming in time. It crosses the main bridge fast, leaving behind a trail of roaring fire. The dark trail it leaves is a meter wide and almost twenty or thirty long. Lots of people have to duck or throw themselves to the ground. When the shadow has passed, they get up and come out into the open. Another rocket has entered where the first opened a hole in the dome.

Dikran is furious. Seeing the plants go up in flames is like watching a slaughter of the innocent.

"I'm going to talk to them."

He leaves the watering hose to Shimbo and motions to Alan and Nicolas to follow him to the control tower where Ivan, Kirill, and other members of the crew are organizing their defenses.

After thirty seconds, Dikran's voice can be heard over the Ark's loudspeakers.

"Who are you? What are you called?"

They get no reply. The pirates are not used to direct questions. They react according to temptation, to provocation, but such blunt questions leave them speechless.

"I said, who are you? What are you called?"

After a few seconds, a yell comes from below, in perfect English.

"We are free looters, they call us 3Dbooters."

"Fine, we aren't a cruise ship. The Green Ark is a scientific project." Then he turns to Kirill, who is miming the number nine with his fingers. "We have nothing of value, but we can exchange more precious things, without anyone having to get hurt. You have nine minutes before the Coast Guard arrives. The choice is yours."

Silence. Then the head of the swarm shouts out to find out the value of the proposal.

"What kind of precious things?"

"We use 3D printers too. We have a database of composition formulas."

Alan grabs Dikran by the neck of his T-shirt. "Are you crazy? You want to swap with them?"

"They are criminals, not idiots. If they realize they can save effort and ammunition, they will accept."

Nicolas has already turned the nanomat on and is connecting it to the catalogue of his compositions. "Here's the cloud address. Let's give them a taste of our merchandise."

Then Ivan calls the Coast Guard; in a few sentences he has apprised them of the situation.

The damage is minimal, nothing the printers can't repair in a few hours of work, except for the trees; they will need more time.

CHAPTER THIRTY-SIX

Nika

Alan wakes with a jump at the end of a nightmare: there had been people laughing, people mocking, but the memory is vague, no who, no where they had been. His heart is still beating hard.

He looks up at the stars accompanying the Ark's path across the Aegean Sea. It is the middle of the night, he has slept badly and there is a lot of time to wait before the sun comes up and he'll be able to play some music, the only thing that will calm him down.

He's scared but he doesn't know why. Pulling over his backpack, he draws out the straw and drinks for ten seconds before resting his hand on Silvia's prominent stomach to feel the double life pulsating inside her.

It is something he has been doing frequently during the night, when they are both relaxed.

The phone's scan app has confirmed the baby is a girl. They are not sure about the name, but they are wavering between Lana (an anagram of his own name) and Veronika, because Silvia likes the shortened version 'Nika', a nod to the winged goddess of victory.

He's worried that Lana-Nika might not be healthy, that there might be complications during the birth, or goodness knows what strange things during the pregnancy caused by the nanites. Those invisible composition engines that had made him feel like the receiver of a miracle, given him his legs back and removed his need to eat in a traditional way, now seem to him like pathogens: incomprehensible for the human senses, intangible, but scattered everywhere, no consciousness, just like infections, unaware, uncaring of any kind of ethics, obeying only the software written in their nuclei.

The nanites, like genes, have been proliferating for years amongst living beings, though not on a human scale. Information about their spread and circulation is in short supply.

How many pregnancies like this have already run their course?

What if Silvia's milk contains nanites, how will their baby develop and grow?

Alan would have liked to talk about all this with Silvia, but he hasn't had the nerve. She must have been thinking about it too, and doesn't want to transmit her anxiety to the baby. He still wants to wake them both, reassure her that everything will go well with the birth, breastfeeding, and weaning. He already feels like a father.

Miriam would not hesitate to offer him words of comfort, not even now she is spending more and more time with Ivan.

No brusque movements come from Silvia's abdomen. The room in there is getting tighter by the day because the baby is growing quickly, perhaps too quickly.

Silvia turns onto her left side. Usually in this position the baby pushes out with her feet, Alan has felt her doing so against the palm of his hand. It is also the position Silvia took to have her photograph taken for her mother, Anna. The short, almost too concise, text with the picture simply said, "Your niece will be born soon. We are all fine, I hope you are too."

By his calculations they must have conceived between November 8 and 9, during the celebrations when the siege on the viaduct came to an end, so there should be sixteen weeks left before the birth, even if the image on the phone shows details of a more advanced stage of development (nose, mouth, and ears already formed) as if her gestation is proceeding at an accelerated pace.

In any case, Silvia and Alan feel ready, like the rest of the Pulldogs, experiencing this event with a mixture of apprehension and happiness.

Nico was moved by the news that in Silvia's once flat and muscly abdomen a baby was growing; when Rafabel told him, his gaze was serene like a mountain hermit. The tears running down his face pooled in his beard, dampening the gray hairs and even quenching the thirst of his skin. In his new condition even tears take on many connotations.

Sitting in the shade of the palm trees as they pass the Cyclades, all the Pulldogs can do is wait for the first enhanced birth in their group. Miriam has composed a number of nanofiber clothes that will adapt to the shape and temperature of the newborn. The twins, Leira and Ariel, between one concert and another put on for the crew, have made rattles to introduce the baby to the joys of rhythm and melody from an early age. Hakim, on the other hand, after introducing the red kite to the Mediterranean habitat, is practicing making animal sounds and shadow animals with his hands in the light of the moon.

Alan rests his ear against Silvia's stomach and hears a rhythm he had never believed could make him so happy.

"Carry on sleeping. Extraordinary days await us."

Silvia's stomach is taut and smooth. Over the last few days of sailing, the baby's growth has suddenly accelerated, so much so Silvia is sure she will give birth quite soon. Any day during the twenty-second week of pregnancy for Silvia is the right day.

Rafabel is holding one hand, and Alan the other.

"She wants to come out. She's ready."

The box by Silvia's stomach holds a series of field surgical instruments Kirill found in the Ark's infirmary: a scalpel, a pair of scissors, a dilator, various kinds of tweezers, a needle holder, a needle threader, and a number of spatulas. Old things, dusty and unused because since the nanites were introduced, most medical stitching and other minor operations have become a thing of the past.

Miriam lifts her gloved fingers. Her hands are shaking.

"Am I supposed to use those instruments? I don't think I can do it on my own. On the World Food Program I did a first aid course, but I'm not a midwife."

Dikran, at the foot of the bed, is looking for a tele-assistance midwifery service.

"Just a minute and I'll find an expert who can follow you step by step."

Lying down, hooked up to the monitoring equipment, Silvia cannot move. Every five seconds she shifts her eyes from Miriam to Alan to Dikran, who seem to have the fate of her delivery in their hands.

"Here we go, Marali Kapoor, twenty-seven, diploma in obstetrics and three years of experience as a midwife."

An unknown woman, going in and out of her own kitchen surrounded by pots and pans, plates and packets of spices, has connected with them from New Delhi on the infirmary's screen. She speaks in Hindi; her voice is translated in real time by a virtual assistant. After the usual introductions the doctor asks to see a scan and Dikran pulls out his smartphone, opens an app and runs it over Silvia's belly.

As soon as the outline of the baby appears on the screen, Marali checks the baby is head down.

"The placenta is posterior and there is a good amount of liquid. Everything is fine. Now we will take a better look to see where we are starting from."

Miriam follows the midwife's instructions and puts her hands where the display shows shadows, and makes the same movements as Marali, who is piloting remotely. After a few seconds, her hands stop shaking.

"Well," says Marali as she occasionally stirs the contents of a big pan, "you are already dilated four or five centimeters."

"I've been like this for about a week already."

"I see, now you are going to feel a little discomfort."

The woman moves her hands as if pulling away the uterus membrane to stimulate contractions.

"We are lucky, you're not going to need the tapes, only a drip."

Silvia begins to be scared, not about the passing pain coming strongly then vanishing and coming back again in steeply climbing waves – she can cope with that – but about the pain that will come soon.

Marali asks Alan to use the instrument for measuring contractions, then Miriam takes a patch for measuring the baby's heartbeat and rests it on Silvia's belly.

Silence.

Miriam moves the patch a few centimeters, searching for Lana-Nika's heartbeat.

Still silence.

Silvia looks at Miriam, who looks at Alan, who is beginning to show his nerves. No heartbeat other than the mother's.

"It was fine earlier. Strange," says the midwife, draining vegetables in an aluminum colander.

Silvia nods, even though she doesn't know what to say. Marali puts the vegetables in a bowl and starts preparing a sauce with tomatoes and spices.

"So…either that thing doesn't work or, seeing as you are already having contractions, maybe the baby is on the way and it's difficult to pick up her heartbeat. Let's try again, so we can see exactly how she's positioned and place the sensor directly over her heart."

Miriam tries to capture Lana-Nika's heartbeat again. And finds silence.

"Wait. I want to talk to the matron."

The midwife leaves the room and after a little while they can hear a bell ringing and someone else speaking in Hindi with her. The virtual assistant can't catch the words to translate them. Silvia and Alan look at each other with terror in their eyes. When she comes back into the kitchen, Marali is smiling.

"Niraja, my neighbor, has worked in the Mahajan hospital for years. She has confirmed that in some cases the sensor doesn't work."

Dikran turns the phone off and on again and this time when Miriam rests the sensor on Silvia's belly, they can hear pulsing in the background, and another stronger, faster one.

Tightening her hand around Alan's, Silvia has another contraction, causing her face to convulse.

"Oh! That hurts! Can someone give me an epidural?"

"You're nearly at seven or eight centimeters. Are you sure?"

"Seven to eight? We're there then!"

"You should start pushing as hard as you can. Judging by where you are, I bet you don't have an anesthetist or an epidural anywhere near you. You'll just have to make the most of doing it how it has been done for thousands of years."

After about five minutes the monitoring equipment shows a slowing of the heartbeat. The baby is about to be born. Silvia grabs Alan's arm, squeezes hard and starts to push.

"Argh! If I live through this, we're going to call her Veronika, promise?"

Alan mimes breathing out; he can't say no to her wishes now, so nods his head.

Then Silvia sees nothing else. She closes her eyes and pushes with all her might.

While a small crowd of people line up to congratulate the mother and cuddle the baby, Alan takes the phone and talks to Marali.

"How is she?"

"Silvia or the baby?"

Alan leaves the room to stand outside in a corner of the corridor because it's so noisy in there. "No riddles please. It's not the right time."

"Silvia is fine, apart from the stress and fatigue from the birth, and I think she is made of strong stuff. A little rest and she will be back on form."

"And the baby?"

On the display he can see Marali laying the table. Her children, carrying backpacks, are running around her sneaking food from the plate of cheese naans.

"Well, the child's condition is...unusual."

"What do you mean?"

"She was born at twenty-two weeks and usually premature babies need extra care as well as a period in an incubator; however, Nika's stage of development is that of a baby born at forty weeks. I might not have a huge amount of experience, but I have never come across anything like this before."

"This...unusualness, is it a good or a bad thing?"

The line of people waiting to congratulate Silvia and Alan trails all the way out of the infirmary. Everyone greets him, they pat him on the shoulder, shake his hand, give him the victory air punch. He can only manage a lukewarm smile in answer, his uncertainty not allowing him to enjoy the moment to the full.

"There aren't any complications right now, actually...Nika is fine, she got an Apgar score of nine, five minutes after being born, it's as if she hasn't come even a little early. But in the future? Who can say.... Like I said, I've never heard of anything like this. I would advise you to take her to a pediatrician as soon as possible."

"We will. Thank you for everything you have done."

Marali presses her hands together at chest level, dips her head and takes her leave.

Relieved, Alan goes back into the infirmary and Nicolas comes over and hugs him in congratulations. Then, as if he has heard the conversation with Marali, he takes Alan by the shoulder in a gesture of reassurance.

"If the development of our anatomy is not normal, why should Nika's be normal? In our condition we have to look for balance differently to before."

Alan doesn't answer and Nicolas remains as enigmatic as ever. They head towards the center of all the attention; in the middle of everyone, they can just see that tiny, pink, wrinkly creature wriggling like a fish out of water. Her cries are strong, her puffy cheeks are rosy and her head is covered with a fine down.

The two men move towards the cot and as Silvia lifts Nika up, the new mother's eyes move between the two men's.

When Alan takes her from Silvia's hands decisively, he pulls her close to his chest and hugs her, making sure she doesn't fall. She is so light, so perfect. Her eyes are closed and still sticky. A little bubble of saliva appears on her dry lips.

Without taking his eyes off Nika, Alan caresses Silvia's arm.

"Do you think she can see me?"

"It's too soon. She can't see, but I can."

Nicolas takes a step backwards and waits his turn.

A reddening sun floods the Green Ark's grassy deck where tree trunks cast striped shadows on the ground. The canopies of the eucalyptus trees three meters above them seem to flutter like immense air balloons anchored to their moorings.

As he walks at a good speed, Alan notices that Nicolas is looking at a pair of splendid *quetzals* with turquoise feathers, pecking at what look like red and yellow chili peppers on a plant in the tropical habitat. When he walks past, Nicolas greets him excitedly.

"Incredible, no reaction, and they are *habanero* peppers, the hottest in the world."

"Is this about the problem with your nose?"

"Yes, the birds are insensitive to capsaicin, they don't have the relative receptors. As you can see, they love those chilies, and at the moment they need the vitamin C and carotene they contain for their feather changing."

"I'm sure you'll find a solution sooner or later."

"All I need is to *see* the molecule that destroyed my sense of smell."

Nicolas notices the bundle on Alan's back. He is about to say something, opens his mouth, but Alan anticipates him and moves on. "See you later, I'm going to carry on with my walk."

Alan is on his third turn around the ship's perimeter. He doesn't want to lose his rhythm or more importantly risk his internal organs overheating. But it is different today because it is the first time he and his little girl have gone out on their own, together. Finding yourself from one day to the next with an unknown living being who takes over your whole day, rewrites all your habits and dictates nonnegotiable conditions, takes a lot of getting used to.

Fortunately, in these first few days of life, Nika has been sleeping a lot, even if for short stretches, and anyway, Silvia has had six months of her growing inside her to get used to the idea, whereas he only met her five days ago.

Alan takes the pouch holding Veronika from his shoulders, and puts the baby down in the grass. He repositions the guitortoise on its strap across his shoulders, and picks out an improvised lullaby.

You are a cloud,
I am the wind;
I will take you up there
and we will watch the sunset.
Come down from the mountain,
little white cloud,
I know you are sleepy,
because you are tired.
Pull up the blankets,
and give me your hands,
sleep soundly, until tomorrow.

As soon as he finishes singing, he realizes Nika has been in the sun for almost five minutes.

He puts her straight back in the baby pouch, a bit worried. Her little face isn't red at all, but he thinks he can see yellow marks on her neck, just visible against the pink of her skin. It could be jaundice, because, according to what Silvia has told him, red blood cell exchange is faster in babies than in adults, and her corpuscles are powerful artificial red blood cells inherited from Alan and Silvia. Apart from that, her liver is developing and perhaps it can't expel the right amount of bilirubin from her blood.

Alan uncovers one of Nika's shoulders and notices it is covered with a just-visible evenly spread rash of dots with defined shapes repeating like changing patterns; sometimes they look geometric, and sometimes...organic. They aren't all the same color either, from ruby red to bronze, to gray. He looks closer, only a few centimeters away, runs his finger over them and, even though he isn't sure, he thinks they are heliotrons.

CHAPTER THIRTY-SEVEN

Fabtotum

The Green Ark heads into the Dardanelles, the long channel that leads from the Aegean to the Sea of Marmara and towards Istanbul. Some ships and a ferry are waiting their turn to slip into the opening; there is a lot of naval traffic.

During their frugal supper of seeds, berries, and fruit, Kirill looks worried. Instead of telling anecdotes about his past, from winter *turnik men* races in the forests of Priozersk to the experimental extraction of oil from plastic waste, he rubs his ears, usually a sign he is nervous about something.

"There will be a lot of controls and then the Bosphorus. This ship is no longer a wreck, how many years has it been?"

"Since 2001, thirty-one years. Are you thinking nano-detectors?"

Kirill lifts his gaze towards the geodesic domes. Until the previous month the tree pollen and mushroom spores, practically invisible to the naked eye, moved around where the wind took them, whereas now, since Nicolas composed heliotrons to test on some of these specimens, the geodesic domes are surrounded by a faint luminescent smudge.

"Our papers are all in order," Ivan says, "the crew are all in order and above board, and the symbol of the United Nations on our hull should guarantee us safe passage."

"I don't know, all they need is an app for tracking pollens and spores. With a little effort they can trace them back to the source of their emission. In some countries, like Turkey, nanites are illegal."

"We cannot avoid the Black Sea, we have to take the risk. Hundreds of walkers come from that area every year, from Russia, from the Caspian, and the Middle East."

"Can't they just go the long way around?"

"Don't be cynical."

From both sides of this oppressive stretch of sea, people run to see the floating apparition that is their ship. Thousands of points of light dot the triangles making up the geodesic domes, making the Ark look like three huge Christmas cakes. When their flotilla reaches the narrowest point at the Çanakkale promontory, passing like part of a parade, everybody watching claps, waving and cheering, almost touching the ship.

The Pulldogs and members of the crew answer enthusiastically. Only Alan notices all this jubilation has attracted the attention of the *Sahil Güvenlik*, the Turkish coast guard. A patrol boat has left the quayside of Çanakkale's harbor and is about to reach them.

The commander of the Harbor Authorities is called Alp Bülak; it says so on the badge pinned to his uniform. In Turkish it means 'warrior of light'. Tall and elegant, just his stare is as penetrating as being searched. The three soldiers of his escort put two bags full of equipment on the ground.

After introducing themselves, Kirill and Ivan explain in English the reasons behind their scientific expedition, and show him the ship's papers. In the meantime, the soldiers proceed to assemble three large spotlights on tripods, each about two meters high.

After looking at them quickly, the commander hands the papers back to Kirill.

"Distributed computing, biodiversity conservation, United Nations, I am impressed," Alp Bülak replies in perfect English. "However, we have to check whether you are knowing or unknowing carriers of illegal nanotechnology. I know it might seem like a pointless waste of time to you, we've already been through all this with GMOs. You'll be thinking there is no point resisting, I know all your arguments. But listen, I am in a complicated position: on the one hand I have to protect my country's identity from possible external contamination, a crusade to spread God's meme through the Turkish people using nanites resulting in the conversion of the Turkish people would be a disaster. On the other hand, being an atheist, I cannot resort to a nanotechnological Jihad, as a number of imams would be willing to in order to beat the infidel enemy. So please, work with me and I guarantee that within twenty minutes we will have finished our stroboscopic scanning."

Alan is watching the scene from behind a bush, hears everything and alerts the rest of the Pulldogs. Paolo and Farisa have seen how the situation is developing too, and now the old man goes to Alan.

"May I ask you a favor?"

Alan has no time for chitchat, but Paolo's air of authority convinces him to stop and listen. "Of course, anything."

"I would like to entrust to you the most precious thing in my life," he says as he pushes an incredulous Farisa in front of him.

"Paolo, what are you doing? Why?"

"Because your work here is done. You brought me to my friends. This is more than I could have hoped for. But your journey must go on."

"What if I want to stay with you?"

He looks at her like a father would look at a difficult child. "We both know our paths will separate anyway, in a year's time? Three? Whereas these people are building a future and I know you can help each other."

Farisa has tears in her eyes. After a last hug with Paolo they leave the main deck, cross the connecting gangway and, within a few minutes, reach the last ship in the flotilla, a barge loaded with earth where willows and olives grow. If the Ark is taken, at a sign from Ivan a sailor will release the red kite which will head straight for Hakim's arm; the signal to leave.

The only thing visible from that distance is the flashing of the stroboscopic lights pinpointing nanites, and light emitted by lasers inhibiting their orientation and aiming systems. A minute later, tarpaulins are pulled over the geodesic domes.

"Why are they covering the trees?" asks Pino as Nicolas picks him up to boost his morale.

"The tarpaulins are made of adhesive materials to block the pollens. They won't hurt the trees, they just want to protect them."

Actually, the trees might well be destroyed. The augmented ones, anyway.

The shape of the red kite emerges from the darkness, gliding silently to land on Hakim's wrist.

"I hope Ivan can do the same one day with my arm," Miriam says.

Tommaso and Kenshij start undoing the cables holding a hanging lifeboat while Nicolas and Dikran do the same with another. The barge's crew give Ariel and Leira two cans of petrol. Shimbo the castaway arrives just as they are about to lower the lifeboat into the sea.

"I know you don't need a cook...but the sea hasn't brought me any luck and I'd like to try on land."

Alan hands him a life jacket. "Jump in."

The lifeboat casts off from the gangplank and with the engine off, the Pulldogs pull away from the Ark using their oars. Miriam is sitting holding tight to the handles, and her air of melancholy is not lost on Alan.

"Did you like the idea of sailing on the Ark?" her son asks her.

Alan guesses at the reasons behind the following silence; he knows Miriam won't talk in front of the others, not now, anyway. He has seen Ivan and her chatting and strolling for hours, as if they weren't on the Ark but in one of those Roman parks where they used to go on weekends when they were seeing each other. He can read it on her face that although she has spent three months with him, it isn't enough.

"Part of me would have stayed, but another part couldn't leave you, Silvia, and little Veronika. I'm a grandmother now."

The next day the monotonous dreamlike song that habitually wakes the camp scattered over the pebble beach is sung by a human voice, instead of the usual Tibetan trumpet. A muezzin who got up before Alan is calling the faithful to morning prayer on the island of Büyükada, an hour and a half from Istanbul.

Covered by branches, the rubber dinghies have been hidden where they landed. It will only be possible to use them if they find some fuel, or continue on oar power.

Ivan called them yesterday, in the middle of the night, telling them the Ark and its following had been subjected to a precautionary seizure while the authorities worked out how to act, a fine or removal from the Turkish coasts. He, Kirill, and the crew have stayed on board and are waiting for a decision to be made. They play long games of chess and take care of the habitats damaged by the pirates. In any case, Alp Bülak is treating them well and has promised to go to see them to assess the situation.

The escaped group's choice of landing point was partially made due to their shortage of fuel, which wasn't enough to reach Istanbul, and partially Kenshij's suggestion. In a few words he made them understand that of all the islands in the prince's archipelago, Büyükada would be the best one to stop on. He united his hands eloquently as if to say he had been here a long time ago.

Now, in the light of day, Alan realizes how right Kenshij had been: behind a stone wall marking the division between the beach and a fallow field full of bushes and purple flowers, some palm trees and red pines loom over ancient Ottoman mansions surrounded by leafy vegetation. Veronika, wrapped in a sheet next to Silvia, is sleeping. Miriam has just opened her eyes, but the others are still resting, except for Rafabel, who must have gone looking for fruit and flowers to put some breakfast together.

Alan decides to take a look around the area.

Going by the number of mansions and sumptuous homes enriching the panorama, the island must be densely populated, but there are no sounds in the area other than Alan's solitary footsteps. He is climbing stairs dug into the side of a hill leading to a deserted asphalted road.

The smells of the Turkish spring explode in his nostrils and he almost, despite their differences, feels sorry for Nicolas, forced to experience this place as a torture of olfactory deprivation. Every ten steps he comes across a different smell: the perfume of lavender flowers, laurel, lemons, and then the balsamic odors of lilies and jasmine.

Alan stops suddenly. He can hear hooves clip-clopping towards him, then a horse-drawn buggy comes into view. The boy driving is waving his hands around as he tells Rafabel the history of the island, as if this is a tourist excursion.

"At the start of the last century the island was a very popular holiday destination for Jews, Greeks, Turks, and Armenians, many of whom had houses here. When the Greeks left Istanbul in the fifties after a wave of violence against ethnic minorities, their homes fell into disrepair. The well-off Turks ignored Büyükada and went to Bodrum in the Mediterranean, and the island has become a destination for the poor of Istanbul looking for entertainment, beach picnics, and horse-and-cart rides."

When Rafabel sees Alan. she waves him up.

"C'mon, get in! It's really worth it!"

She is sitting on an embroidered cushion, her hands, held together in her lap, are painted with dark henna and she has a necklace of silver bells.

"Over the last few years Büyükada has come back to life," the guide continues. "Some ancient families from Istanbul have started coming back to their summer homes, investors have remodeled old properties, and many artists, writers, and foreigners come here to rest, find respite from the modern world, and live peacefully in a kind of Arcadia."

Older couples with veils are sitting on the crest of the hill overlooking the blue of the Marmara Sea.

"It's like going back in time, just two hours from the chaos of the Bosphorus."

Cascades of wisteria blossoms tickle their heads as they pass ornate mansions with ornamental terraces, buttresses and fancy shutters. Some have just been remodeled, others look more fragile, paint peeling, wooden beams in full view and ruined shutters hanging lopsidedly on their hinges. Abandoned buildings that have absorbed so much history as to command respect, almost veneration.

"This is definitely a place where you could stop and raise a family," says Rafabel dreamily.

"It's such a pity that every Garden of Eden has its own snake, but I imagine you're not saying it just because," Alan says, turning towards her.

"Of course Veronika is a very precious gift, but she will need care and attention like any newborn baby."

"Are you insinuating she doesn't have all she needs?"

"Perhaps the cold of the northern forests isn't the best thing for such a small baby."

"I believe babies have been born and raised in Russia for quite a long time. In our condition the cold temperatures would actually help us avoid getting fevers, from overheating. I have thought about it, Rafabel, I am not being reckless."

"I get it, but without a house? Living in forests and by rivers? It reminds me of a story I read years ago. Have you ever heard of Agafia Lykova?"

"No, what has she got to do with anything?"

"Well, Agafia's family, Orthodox Russians, moved to the Siberian taiga in 1936 to escape persecution and protect their lifestyle. After a long voyage the Lykovs settled in the Sayan mountains. Doesn't that mean anything to you?"

"I get the analogy, and?"

"Well, during their isolation two children were born; they ended up living in a hovel by the Abakan River, hundreds of kilometers from the nearest villages. In 1978 a helicopter pilot who was flying over the area during a geological research mission found them by chance. When the geologists made contact with the family, the Lykovs decided not to leave their home. When her parents died, Agafia, the last remaining Lykov, decided to stay there, living in solitude."

"Is she still alive?"

"I think so. She must be at least a hundred."

"If I get that far I'll go and visit her. You see danger where I see salvation. And what's more, this Agafia could be just the thing for

Nicolas. She sounds like she would be in harmony with his mission of cultural diversity."

"I don't like how you treat him. And from the way you talk, it sounds as though you are expecting yet another fight."

"Why, aren't you?" Alan says scornfully.

"I hope you find a way to make peace. It was no fun seeing you fighting like that."

The young man pulls on the reins to stop the buggy in front of Hagia Yorgi Orthodox Church. A number of supplicants who want to have state-funded homes have made houses carved out of sugar cubes and stuck them with wax on the darkened votive wall, and there are various street sellers of all kinds of 3D-printed junk, from jewelry to underwear. The guide wants to say something, but Alan and Rafabel's discussion has reached a point where they are raising their voices.

"Have you forgotten that *it was me* who founded the viaduct?"

"You're forgetting our community existed *before you did that*, at Serra Spino."

"And the nanites? I brought them to the community."

"OK then, let me see if this is what you are saying, you freed us from the need for food so we would become your slaves?"

Alan looks the other way and carries on talking as he looks at the view.

"No, but I saved the viaduct from speculators and that shows I accept my responsibilities," he says, but can't meet Rafabel's eyes. "He, on the other hand, wants to be given greater status, on what grounds? Because of the infusions and perfumes he composes? Because of the heliotrons?" He turns back to face her. "Think about it, Rafabel, what has Nicolas done for the Pulldogs that hasn't also been for himself?"

"Well, he has...."

This time it is she who lowers her eyes and accepts the blow. The young man, seeing the situation, decides to start the horse walking again and Alan continues to push his point.

"How can we know whether his choices are the right ones?"

"And yours? Isn't the principle the same?"

"We'll just have to wait till the next time to see how it works out."

"The next time?"

"Yeah, we haven't finished this discussion. You'll see…. Nicolas might be creative, but he's predictable. It is him who picks the fights. It is him who is no longer satisfied. There will be another dispute as soon as the occasion arises."

"Do you mean about our destination?"

"Exactly, Russia or the rest of the world? We all know whose side you'll be on."

"As far as that goes your position is crystal clear too."

"You are a minority. You, Nicolas, Hakim and Kenshij, all of you who took the heliotrons. Deal with it."

They both look away, both in a sulk. After thirty seconds the young driver stops the horse in İskele Meydanı, the island's main piazza, by a Belle Époque-style clock tower which, instead of having a cuckoo, has two statues: an imam who appears on the balcony at prayer time and says, "God is great!" three times, and then a man without a moustache who comes out on the hour and says, "How wonderful to be Turkish, Turkish, Turkish!"

The boy says, "The tour is over."

"Jasmine! Yes, it is! It's back at last!"

Nicolas is jumping around a bag of dried flowers, waving his phone in the air. Alan goes over to him and tries to grab his arm to hold him back.

"What's the matter this time?"

"I can smell again! A stall holder gave me these jasmine flowers this morning."

"You look like you've gone mad."

"Mad? I was mad before when I couldn't smell. I'm sane again now…look at this."

On the screen there is a map of Büyükada island, but instead of being labeled by place names, areas of odors are mapped out. A key by the side lists the various categories:

Aromatic: camphor, lavender, menthol, bay, lemon
Balsamic: vanilla, lily, jasmine
Garlicky: thiol, amine, rotten eggs, bromide
Empyreumatic: coffee, toasted bread, tobacco smoke, tar, mothballs, petrol
Goaty: cheese, sweat, urine
Repellent or suffocating: pepper, coriander, orchids
Nauseating: rotting meat, indole, excrement, stapelia (asclepiadaceae)

There's a mixture of jasmine and lavender hanging around their encampment, with an undertone of sweat and alcohol; closer to the coast the stink of cigarettes creeps in, mingling with the aggressive perfumes of the girls in the bathing resorts, weakly diluted by suncream.

"How did you manage to map the island's smells?"

"Actually, at the moment all I can smell is jasmine. The other smells are graphic extrapolations based on the local plants and above all real time information from the net and social networks, photos, videos… y'know, Big Data."

"How did that happen?"

"IFT88! That was the protein receptor that damaged the cilia in my nose. I programmed a nanite with the normal gene sequence of the IFT88 protein which installed itself on the cells' organelles via a simple common cold virus. Then I monitored the growth of the cilia amongst the nerve cells. After seven days they started growing back, extending the cilia beyond the dendrites, which is what the olfactory neurons need to detect smells."

"And where did you find a *normal* sequence?"

Nicolas points at Hakim, sitting stroking the red kite. He smiles, wrinkling his nose and flaring his nostrils.

"Anyone could have given a pinch of olfactory cells," Nicolas answers. "But he offered. Anyway, it will take some time before I can smell things like before. In the meantime, I can practice here on the island. With this app it will be easy."

"Ah, this is certainly the right place for learning to smell."

This stops Nicolas in his tracks. "My mum used to say that about the Rendezvous."

"I'm sorry, I didn't mean to open old wounds."

"No worries. She had nothing to do with it. Actually, I think it's about time, after so many months, to write to her to tell her my good news."

"Won't your father find out? Won't he work out who knows what new torture to punish you?"

"Maybe you're right. I'd better not mention my nose, I'll just tell her I'm doing fine."

Alan wanders off and greets Silvia, who is holding Veronika, eyes still closed and wrapped in a silvery blanket, in her arms.

"Good morning. How are you both?"

Silvia lifts her shirt and begins to feed the baby.

"Never better. How is the island?"

"A paradise of Ottoman mansions, beaches, and lush, verdant hills... ah, no cars, only horse-drawn carts. I took a tour with Rafabel."

"So then, can we stay here for a few days?"

The recent argument comes back into his mind. Alan is about to answer when Nicolas appears behind him, armed with his bandoliers and magic pen.

"I composed this for Veronika this morning. Can I hold her for a bit?"

He is holding a pacifier made out of an elastic substance, like rubber.

Alan and Silvia look into each other's eyes, she is the one who answers.

"In general we don't want to give her a pacifier, but once in a while for a little bit is all right. Please though, stay close by, where we can see you."

Nicolas takes the wrapped bundle that is Veronika and takes a few steps, looking for a place where flowers are growing. He stops by a bush of Italian strawflower and bright white viburnum. Then he settles the infant on the grass and sits on a tree stump. He unbuckles his bandolier, connects the pen, and starts telling a story.

★ ★ ★

After ten minutes watching from a distance, Alan makes his way over to Veronika and Nicolas, who is moving the pen around in all directions. Every now and then he connects the oblong reservoir to a different cartridge from his belt to change the compositional material the *fabtotum* 3D printer is using. As soon as the filament coming out solidifies, it shapes part of an animal forming before their eyes.

"Can you tell me which way I have to go, please?" says 'Alice' in Nicolas's falsetto voice.

"It depends where you want to go really," replies the Cheshire cat, this time in Nicolas's boy voice. The cat's whiskers are white, its fur is brown with ginger stripes.

"It's not very important where…" Alice continues.

Nicolas puts the tail on the back of the cat and wags it from left to right.

"So any road is a good one," concludes the animal.

In that moment Veronika opens her eyes; they are as bright as a bird of prey's, taking in everything happening around her. Her eyes are warm brown with golden flashes. She doesn't just look around her, she is looking for and finding information.

The first face she sees is Nicolas's.

Alan sees this and runs quickly to put Veronika back in her sheet. He shoots a nervous look at Nicolas and then runs to Silvia. As soon as he reaches her, his breathlessness is not so much in his breathing but in his hands shaking with suspicion. He grabs her by the shoulder, almost pulling her.

"I have to talk to you. It's urgent."

Silvia stops arranging the flowers she is getting ready for lunch and follows him. Far from the eyes of the others, Alan turns Veronika around, lifts her T-shirt and lets the sun shine on her.

"Don't tell me you haven't noticed."

The reaction isn't immediate. Silvia doesn't answer.

"Did you take heliotrons without telling me?"

The dots are beginning to appear on the child's skin as Silvia shakes her head.

"You know it is the only way Veronika could have them, right?"

Silvia continues shaking her head.

"I am asking you, clearly. Is Veronika his daughter?"

She stares at him and deliberately over-articulates every word. "Alan-you-are-the-father."

"So where did these things come from?" he says, pointing at Veronika's skin. He hurries to cover her as if she has some kind of psoriasis to be ashamed of.

"I don't know. She could have got them some other way. Do you always know how you get ill? Those things are nanotechnology too, right?"

Silvia leans over to Alan and pulls him to her hard. She tries to kiss him but he wriggles out of it, and bites the hollow of her neck to release some of his anger.

"Maybe, maybe not, who knows, perhaps for Veronika you wanted an intelligent father like Nicolas."

Silvia punches him in the ribs. "Alan! What the fuck are you saying? Don't start playing the victim."

"Someone who creates perfumes, a successful designer who always has fantastic ideas about making the world a better place and can compose all kinds of objects to make the wishes of a baby come true.... Tell me that's not how it is."

"That's not how it is. I'll say it again. For me, you are her father."

Before he can answer, Silvia covers his mouth with her hand, and then her lips.

CHAPTER THIRTY-EIGHT

Prinkipo Orphanage

It has been five days and they have not heard from Ivan yet, a sign that the Ark might not be able to set sail anytime soon. Alan wants to leave and head north, but he knows that right now he would encounter a dangerous amount of resistance. Not least because the summer season has started and during the weekends Büyükada is overrun by hordes of tourists from all over the world.

After passing a five-star hotel – the Prenses Koyu – with white deck chairs and a luxury patio, Alan decides to go down to the green carpet-covered cement jetty. Hundreds of bathers are hanging out under striped beach umbrellas, while waiters run here and there bringing drinks and narghilè with smartfumes to inhale.

His talk with Silvia has not removed his doubts about Veronika's paternity, and the discussion with Rafabel has left deep wakes in his mood. Sooner or later his vision of the Pulldogs' future will be questioned, so he has to make it more solid and appear more palatable, otherwise Nicolas's faction might manage to impose a different destination, or as far as he can make out, an undefined destination, intangible, a kind of *eternal wandering* without a precise destination or stages.

A band of clowns in swimming costumes wave and grin in front of the phones of dozens of mums wearing sarongs and sunglasses. Their children are spraying each other with water rifles, diving down spiral slides and playing on an inflatable raft ten meters from the beach, watched over by supervisors dressed up as sea monsters and cartoon characters.

Alan takes his daughter into his arms, puts the backpack on the ground and climbs down the ladder from the jetty into the water. Veronika lets out a squeal halfway between surprise and pleasure. Her familiarity with water means she floats naturally, and when her swimming cap gets wet she finds it even more fun.

They have been splashing around for a minute when Silvia appears from the bushes behind the beach. At first she walks at a normal pace, then she starts running through the deck chairs to the jetty and dives into the sea. Everybody turns to watch her with the same uninvolved surprise they would reserve for an unexpected stain on a swimming costume.

"You did that on purpose. Do you want to ruin their holidays?" Alan pulls her between him and Veronika.

"Bah, holidays.... These aren't the people who have a short break and crowd the beaches at the weekend to recover from the working week. These are Istanbul's upper bourgeois, and exotic global holidaymakers. People who move from one state of immobility to another; from the office and car in the city to the beach and a resort hotel room. They aren't holidaymakers, they are 'holidayed'."

They leave Veronika to float. Every so often Alan gently pulls her foot or hand to bring her closer.

"Don't judge. I can't find it in me to be annoyed with them if they want to use their time to the full, take advantage of the occasion to see a place like this and make the most of their holidays."

Veronika's skin is beginning to redden, and the dots that were at first only visible on her back and chest have now spread to her neck, arms, and legs. Silvia notices and goes back into mum mode. "Shall we go back to the camp?"

The three of them climb back up the ladder to the jetty and leave the beach behind. They don't take the pretty tree-lined avenue Tasikar Yolu, preferring the footpath crossing the southern side of the island. The sun is filtering through the tree branches, creating shafts of light that Alan and Silvia have fun dodging, making Veronika laugh. Her cries of happiness continue for a dozen or so minutes until Alan lifts his eyes and sees Tommaso and Pilar perched on a branch, the eternal lovers locked in a kiss.

In embarrassment the young man grabs a branch, and, losing his grip and balance at the same time, he falls five meters with a horrendous crash, and the sound of bones breaking. "Ah.... My knee!" Tommaso yells. "It's broken!"

The concave area between his thigh and calf is now convex. His kneecap must have broken in various places and the whole lot, femur, tibia, and fibula, are done.

Pilar clambers down the tree as fast as a monkey and runs to his aid. "We have to take him to a hospital."

"I don't know if there even is one here," says Alan. "Don't move him, I'll run to Prenses Koyu."

After crossing a woodland floor full of wildflowers, Pilar clears the path by moving twigs and branches out of the way until they reach the second-highest point of the island, Isa Tepesi.

"This way, we came by here a few days ago. It's impossible not to see it."

Silvia has gone to their camp with Veronika to let the others know about Tommaso's fall, while Alan, after walking back from Prenses Koyu, has lifted the young man onto his shoulders. Even though it has been put in a provisional splint using two branches and some twisted twine, Tommaso yells every time his damaged leg swings and his shouts scare the horses pulling the carriages.

The drivers slow down to let them pass and shout, cussing and encouraging at the same time.

"At the hotel," says Alan, "they said the only hospital on the island is too small. For an operation of this kind, patients are taken by ambulance boat to Kartal, on another island, or to Istanbul. At Prinkipo they should have both the infrastructure and the personnel, but it's private."

At the top of the hill, fenced in, is a four-floor wooden building over thirty meters high that looks like an ancient manor house.

"Actually, they told me it isn't a clinic but a research institute...the bloke on reception didn't really know a lot."

There is no one to greet them at the main gates, so they just go in, crossing a huge lawn with flowerbeds and tree-lined paths. The glass entrance door opens as they reach it and they find themselves faced with a nano-detector. Scanning is obligatory. A series of light and acoustic indicators show the presence of the nanotechnology in their bodies.

Pilar looks at Alan suspiciously. Tommaso is trying hard not to yell, gritting his teeth.

"Now what?"

The place is incredible. The liberty-style ceilings blend with Ottoman architecture and walls covered with climbing plants. The smell of mimosa purifies the air and the floor is covered by a carpet of pine needles crunching under their feet. If Dikran had been there he would have trawled Google for any and all references to the building.

A woman dressed in a suit comes out of a side door. She doesn't look like a nurse or a hostess, she looks serious and distinguished, like a researcher: her black hair is tied in a bun at the nape of her neck and she is wearing overlarge glasses made that way to accommodate their AR functions.

Noticing the new arrivals are not locals, she speaks to them in English rather than Turkish. "In normal circumstances we only receive invited guests. Your presence is rather…unusual. I am Esin Demir, how can I help you?"

"Our friend has had an accident. He fell and smashed his knee on a rock. At the hotel they told me to bring him here."

The woman raises an eyebrow and puts on a bland smile. "This is not a hospital."

"Yes, they told us, but can you help us? Do you have an infirmary where you can check him out? Then we will look for an alternative."

The woman looks at the strangers more carefully. "You don't look like tourists. So what are you doing here, if I might ask?"

Alan dumps Tommaso on a sofa and Pilar answers with all the sincerity her mouth can reveal.

"We were on board the Green Ark, but the harbor authorities detained it because…."

She feels someone pulling her arm. Alan darts her a stern look, but for Esin Demir this information is anything but negative.

"Oh, I know all about the Green Ark. The society I represent is one of the project's financiers, we have contributed to its composition by donating five years of work over our servers."

Ms. Demir taps on one of the arms of her spectacles and makes a call. "Please, send a stretcher to the entrance hall. Pass me Doctor Çakmak too, it's urgent."

Alan cautions her, to avoid any misunderstanding. "We can't pay. We don't have any money."

"Oh, that won't be a problem. I'm sure we can come to an agreement, given the results of the entrance scans."

Two orderlies wearing white arrive looking more like uniformed waiters. They lift Tommaso onto the levitating stretcher. The slight stiffness of their movements betrays the artificial nature of these orderlies.

Ms. Demir motions them to follow her. "Please, this way."

Alan shoots an irritated look at Pilar. She sidles closer to him and whispers, "But isn't nanotechnology illegal in Turkey?"

He shrugs.

"Alexandre Vallaury, a Turkish-French architect, designed what later became the Greek Orthodox orphanage of Prinkipo at the height of the Belle Époque. It started life as a hotel casino for the International Wagon-Lits Company, the European company running the Orient Express," says Ms. Demir as she accompanies Alan and Pilar back to the entrance hall.

"When Sultan Abdul Hamid II forbade gambling the building was bought by the wife of a Greek investor who donated it to the Ecumenical Patriarchate of Constantinople, who converted it into an orphanage. At the height of its activity, it could house up to a thousand children. Unfortunately, it was closed in 1964 and since then it has been used for various things and abandoned dozens of times. Then there was a terrible fire in the eighties which compromised its stability."

Tommaso has been checked over. Doctor Çakmak found a triple fracture of the knee; fortunately the ligaments are undamaged. After twenty minutes Tommaso was taken to the operating room where Doctor Çakmak is wearing a pair of Ryu 43 gloves for AI-assisted surgery.

"Surrounded by a thick forest of pine trees, these twenty thousand square meters are the largest historic wooden building in Europe. In 2010 a long dispute between the directors of the Turkish Foundation and the Ecumenical Patriarchate was resolved by the European Court of Human Rights, which ordered the building to be returned to the Greek Patriarchate. In 2021, following the Greek troubles, it was sold to a Swedish investment fund which I represent."

At the end of the operation cleaning the damaged area of bone fragments and reconstructing the kneecap, the doctors apply a drain and a brace to keep the joint solid for the first weeks of healing. To attenuate the pain, he has been fitted with a biodegradable microchip in his calf which releases painkillers and anti-inflammatory drugs during the post-procedure period.

"Seven years ago, the New York architect Nicholas Koutsomitis redesigned the orphanage to become an international research center financed by some of the most innovative companies in the world. From the outside it is extraordinary," Demir says, "though they say that inside it is haunted by ghosts of some of the unfortunate orphans."

From the windows on the north side of the building they can see a futuristic tent city powered by the sun's rays. There must be hundreds of tents, not so tightly packed and each one different to the next.

"As you can well imagine, our interests embrace all the convergent technologies, Nano-Bio-Info-Cognitive, in brief, NBIC."

Alan isn't receiving danger signals from Ms. Demir's behavior; all the same, he is staying on the defensive.

"Thank you, but I still haven't worked out what you are asking of us."

"As far as I understand it, there are about twenty of you camped on the beach." Alan nods. "So then, why don't you come and stay here? You can compose your tents next to the others and use the creature comforts available to the researchers. In exchange we would like to be able to examine your anatomy. You are one of the first cases of man-nanite symbiosis we have come across. We would be truly grateful."

The rest of the Pulldogs are waiting outside the entrance for news of Tommaso.

"I hope to see you again soon. I mean it...." Esin Demir waves and goes back into the building.

A team of android orderlies are carrying silver platters of grilled vegetables arranged like mosaics on their shoulders: beautiful salads of green beans, butter beans, and spiced tomatoes, and pyramids of hot *börek*, triangles of pasta filled with cheese. The narghilès, loaded with relaxing essences for inhaling, have been placed in the centers of the low tables.

From the camp it is possible look north and see giant cargo ships anchored in the Bosphorus area. The Green Ark must be down there somewhere: the order revoking their detention arrived yesterday and Ivan called them to say how impatient Kirill was to set sail for the Black Sea and Yalta. If the Pulldogs intend to board and continue on the Ark they will have to hurry to join them, or else wait for them to come back in a couple of months.

Alan and Silvia are inhaling from the same narghilè, as Miriam has taken Veronika for a walk. Life on the hill is pleasant, with relaxed rhythms; even Nicolas seems to have become calmer now he has regained the power to smell. He shares a tent with Rafabel now, as she has allowed Tasia and Pino to compose their own teepee where they can play and sleep together.

Tommaso and Pilar, who have only ever had one tent, run past Alan and Silvia giggling after being in the massage room. She's wearing a sarong, he a pair of Bermuda shorts, and like every day they will soon be heading for the swimming pool.

"I think his knee is fine, even though it has only been two weeks," says Silvia.

When he walks, Tommaso can put all his weight on the damaged leg without any problems. Five days ago he stopped using crutches. After waving at them he slips into the tent with Pilar.

"Just look at him, he is fine, even without physiotherapy."

Miriam appears from the clearing with Veronika in the baby pouch and lets Silvia know her baby is hungry.

"We have no excuses left," says Alan. "We have to start walking again...."

He has barely finished speaking when Valeria appears behind him.

"Do you realize this is the first restaurant we have been in since we left?"

"Restaurant? I can see people standing, lying, picking at fried sweets, nibbling at nuts, inhaling smartfumes, people are having massages, swimming and listening to music."

Right from their first day here, Prinkipo has been a difficult place for Alan to define. Officially – according to what Esin Demir and Doctor Çakmak say – the orphanage is a research center; in practice, though, it seems to be a holiday resort for wandering geniuses.

"True, but the meals and the service are restaurant standard. You tell me, is this the world we left behind us?"

Valeria raises her glass. Alan doesn't react, so she clinks it against their narghilé.

"Well...here's to the world we left behind us."

When she has gone Alan inhales and exhales heavily, as if trying to blow away her words. Silvia does nothing to encourage him.

"If this is any indication of the general feeling, I doubt anyone is going to want to leave."

Miriam hands Veronika back to her mother, who starts nursing her. As soon as the infant latches on, Silvia lets out a sigh, almost without realizing it.

"What's the matter?" Alan asks her.

"Nothing."

"You sighed."

"I was thinking about Büyükada's perennial summer, the softness of life here at Prinkipo. How easy it would be to stay."

"Yeah, right." Alan says sarcastically. "Forever on this beach. Always in the sun, lying by the sea. So easy."

Then Silvia points to the twins. Their tent is a kind of permanent gathering place for musicians; they jam and experiment with all kinds of musical instruments.

"I think you might have lost your percussionists, too. It's your choice, which group, the players or the walkers?"

Alan doesn't speak as the Pulldogs swelter in the sun and have fun with the researchers, laughing, drinking and inhaling. When Veronika is full, Alan takes his daughter in his arms and decides to go for a walk.

"I'll burp her. This evening I'll have to talk to everyone in a general meeting. I don't like all this. It's not like we're on the island of Ogygia."

CHAPTER THIRTY-NINE

Family Roots

"Please, sit down, Alan. How are you enjoying your stay with us here at Prinkipo?"

Doctor Çakmak's office is on the fourth floor of the orphanage: a clean room with a yellow mechanical arm with chrome finishings hanging over an operating table. The doctor's gloves are stored in a showcase against the wall.

"It's all so different to what we are used to, I didn't realize the Pulldogs missed their creature comforts so much...."

After a meal of milk enriched with nanites, Veronika is sleeping in her pouch on Alan's chest.

"It can't be easy to live like you all do. Exposed to the elements and, what's worse, the prejudices of strangers, and now you have an infant, too."

"Perhaps at the start it was difficult, but you get used to it."

"You wanted to talk to me about something? Is Tommaso doing well?"

"Yes, really well. Actually, I wanted to thank you for everything you have done."

"I am the one thanking you. Your arrival has given new impetus to our research. Those nanites are revolutionary, and when I think they were composed six years ago by an unknown author, well, it gives me goose bumps."

"I couldn't believe it either. Then, when I managed to move a toe, I had to believe."

"Perhaps we should try to contact him."

"I don't think the Master wants to be found."

"It's a pity...I would have liked to have invited him here."

"Why are you so interested in the nanites?"

"On one hand, it's my job, even though I now limit myself to supervising the procedures executed by the AIs, but on the other hand, I have a personal motive. Ten years ago my wife discovered she had breast cancer. Have you ever heard about the lawsuit against Myriad?"

Alan shakes his head.

"Well, these madmen wanted to put a patent on the gene causing breast cancer so they could sell the screening test. My wife, together with other women from the international committee put together to fight this absurdity, rightly claimed that the gene, as a product of nature rather than a human invention, could not come under patent law. Luckily, at least for now, the American Supreme Court ruled against Myriad's patent. However, these things never end with a ruling. If Myriad one day obtains the exclusive commercial use rights for the human breast cancer gene, they will make the screening test expensive, and poor women will not be able to afford it. Well, I have developed a nanite, it's not as complex as the one your Master created, which synthesizes that gene."

"I get it, I hope we will still be useful to you when we have gone away."

"Why so soon? Has something happened?"

"No, but if we walkers do not walk, we go back to just being 'enhanced people', nothing special, you could find plenty others like us around, people who have decided to leave the big cities as a kind of rebellion against the world order, or people who only do it for a short period: like the *rainbow voyagers*, mini-mobile-home travelers, sleeping bag people searching for illumination, tramps and otherworlders avoiding the West; there is a flow of people moving continually, people leaving with their families, running away from their families, trying to vanish, trying to change identity. Actually, our goals are larger than this."

"In what way?"

Alan tells the doctor about his vision of a migratory lifestyle spent in amazing places, moving with the seasons like those birds who from one part of the year to the next build new nests which they never live in permanently.

The doctor listens until Veronika wriggles and cries. Alan takes her out of the pouch to rock her in his arms.

"In this world you won't need an identity, there won't be borders to cross, or resources to manage because the sun will power the nanomats, and knowledge and experience will be shared over the web, not sold. If we can do it, we will be the weeds infesting the planet."

"What an enchanting vision...."

"Indeed. However, I came for a different reason."

"I'm sorry, your presence has really shaken up our research."

Alan moves the silvery sheet wrapped around his daughter and shows Veronika to the doctor.

"Had you noticed that some of us have developed this nanite modification?"

"Yes, I have met Nicolas. We are analyzing his heliotrons."

"And has he told you about his project?"

"No, actually he hasn't. He is very reserved."

"Well, the point is I never took the heliotrons."

The doctor looks at Alan in silence.

"I want to know whether Veronika is my daughter or not."

Çakmak stands to open the door of a locker behind him.

"Are you sure? I imagine you have already spoken to Silvia about this."

Alan nods, and the doctor hands him two cotton swabs.

"I need samples of cells from the insides of your and your daughter's cheeks."

"How long will it take?"

"I will have to send the samples to the clinic where I work in Istanbul. I can ask a favor of a colleague, but it will take at least three days."

Alan puts Veronika and her pouch back on his shoulders and gets up.

"Great."

CHAPTER FORTY

Non-Promised Land

Alan unrolls a map of the Mediterranean on the ground and anchors the corners with stones to stop the wind getting at it. With the flashlight on his phone, he points out Rome and traces the route they took to the Mount Adone Animal Rescue Center, then their route to the Adriatic Sea where they boarded the Green Ark.

The beam of light continues along the route taken by the flotilla, crossing the Naval Blockade, the pirate island of Antikythera and their arrival in the Sea of Marmara and on Büyükada. Every so often Alan raises his eyes to look at each of his companions with the control of a general briefing his high command.

"The Hebrew Hag," says Alan, "and the pre-Islamic Haj were journeys, and celebrations in sacred locations with connections to the original nomadic route, the Easter. Before becoming the commemoration of the exodus of Moses's people leaving Egypt, Easter was a commemoration of nomadic conditions and involved a three-day trip in the desert followed by a feast."

The light trailing across the map has brought back memories, reminded them of arguments and rekindled bonds, leaving each of the Pulldogs with joys and disappointments: this route is their life now, it is what unites them, what makes them a family, it is their non-stationary home, their *desti-nation*.

"It is the original character of pilgrimages, the gathering of groups who usually live separately, to celebrate a common bond, that connects the primitive condition of humanity to religious pilgrimages and continues to do so today. A character so deep and irreplaceable that in modern times manifests as musical, football, consumer, and touristic pilgrimages."

Alan switches the flashlight off and starts walking around the circle of Pulldogs.

"But we are not pilgrims, nor fans, not consumers, nor tourists. We do not gather thanks to international transport systems, we walk; we do not identify with mass production, we are the mass that produces for itself; we don't use mass distribution which gives everyone identical things wherever you are, we exchange compositions, knowledge and personal experiences for free, because buying creates hierarchy, whereas donation creates equity."

On his second lap, he stops.

"The next stage in this world, now, is in front of us."

Alan turns his flashlight back on and shines it on a point on the map, Cappadocia in Turkey.

"I don't want to offend anyone, but in just two weeks here at the orphanage we have gone from being walkers to holidaymakers, happy prisoners measuring our cells, toing and froing from the pool to the massage center. Our freedom has become the length of our chains. Will you be able to leave this place?"

There are murmurs and whispers but no one speaks, neither to disagree nor answer. Alan continues, "There is a time for everything, and now the time for resting is over. If any of you are happy with life here, please, we are not going to force anyone to continue, but I have no intention of staying here. Stability is the beginning of the end, you don't need me to tell you, but I want to remind you of one thing: when we walk we can only fall forward."

Alan turns the flashlight off and sits down. "Tomorrow we are setting off again."

He likes the idea of pushing his limits, of always inventing a new obstacle to overcome, a challenge to win.

"I won't ask for a show of hands as if we were in a residents' meeting. Sleep on it and decide in all serenity. We are and always will be Pulldogs, with mesh networks we will always be in touch with each other. All paths split and come together again in the end."

The group says nothing, then, one by one the Pulldogs get up and leave.

At the end of the speech, Nicolas and Hakim go over to Alan by the communal tent.

"We are with you. We have to move as soon as possible. This place is fantastic, Esin and Doctor Çakmak are so kind, but we are at risk of losing sight of our goals."

Alan walks towards the woods surrounding Isa Tepesi hill.

"Talking about goals. Are you still bent on taking nanites to the natives?"

Hakim grabs Alan's arm roughly. "Please, don't call them natives. We are all natives of the place we are born in. It would be better to call them tribes.... A bit like us."

"All right. The problem is they are dying out. Don't you think if there was something useful in their way of life things would be going better for them?"

"The reasons behind their difficulties have nothing to do with their culture," Nicolas says. "If you take away the land where they have lived for thousands of years, if you impose values that have nothing to do with them, if you steal the resources they need to survive, well, it is not the fault of anyone's lifestyle."

"It reminds me of the viaduct."

"Exactly, we defended it with all the means we could."

"But then we left it, and came all this way. And it was you who convinced me no promised land exists; that the concept of place and belonging have to adapt to the times we are living in. But, after thinking about it a lot, I believe you should try. If anyone is capable of pollinating the world with ideas, it's you, Nicolas. I hope you manage to save lots of tribes."

"Why are you saying this? You...I mean, aren't you coming?"

Alan surprises Nicolas by resting a hand on his shoulder.

"This is your path. I don't feel I can lead the Pulldogs into such dangerous places. Now we have Veronika too, I have to look after her, and I hope that in the future other children will come along."

"Are you scared? There are dangers everywhere, and that doesn't stop children being there too."

"Yes, but I have learned to be cautious, it's not the same thing as being scared. Maybe one day you too will understand this."

The morning after, as Hamidiye's mosque's muezzin calls the faithful to worship, the Pulldogs begin to decompose their tents, except for two; the twins' tent is still there, and they are sitting in the opening, sad but determined to stay on the island.

Alan goes over to them, strumming his guitortoise to disperse any tension.

"So am I going to have to post an ad: looking for expert percussionists with no fixed abode?"

"Sorry, Alan, but we're not in the mood for joking," says Ariel, lighting the narghilè. Her sister Leira is in a similarly bad mood.

"I know it wasn't an easy decision. Anyway, you will always know where we are; you just have to look at iMaps."

"It's not the same thing." Leira has tears in her eyes.

"You don't want me to try to make you change your minds, do you? So come and say goodbye, and then come back and play. I want to hear your rhythm all the way to the Anatolian coast."

The women get up and hug all the others.

The other tent that hasn't been decomposed belongs to Martina and her son Pietro. Inside, a separation within the separation is happening. She has found the ideal place to produce her aerography creations, whereas he wants to go with Alan, carry on playing bass in the group, stay with the Pulldogs.

"I know you will be in good hands. At twenty-four you can decide on your own future," Martina says, sliding out of her son's arms. "Now go, I don't want to see you for a while. But contact me every now and then, and come back to see me as soon as you can."

Pietro shrugs on his rucksack, the neck of the bass sticking out of the top. His eyes are wet with tears, but he doesn't want her to see. Alan puts an arm around his shoulders and pulls him away to remove him from his embarrassment. "C'mon, run to the others."

Esin Demir and Doctor Çakmak and Prinkipo's research staff have got up early so they can say goodbye to the Pulldogs. Alan thanks them all for their hospitality and takes advantage of the occasion to give his vision a name.

"This meeting was destined to happen. You have made Tommaso whole, we are leaving Ariel, Leira, and Martina here with you. I'm not sure who has gained more." Someone starts laughing, Rafabel is overcome with emotion. "And after talking to Doctor Çakmak, my ideas about what we can become are much clearer. You see...in my mind, the life we have been leading over the past few months has a name I never mentioned, transhumance."

As soon as he says the word, people start laughing. Alan is happy to play along.

"Yeah, I know, it's because I have origins in Abruzzo, I'm no good as a copywriter." He looks at Miriam with the hint of an ironic smile. "But the doctor suggested something better, the rhizomance. The rhizome is a root that spreads through earth horizontally rather than vertically. A rhizome doesn't have a start or an end, it's somewhere in the middle, imperceptible, it doesn't put down roots. Therefore, our experience has been and will continue to be a rhizomatic transhumance."

The twins' fingers start moving on their bongos. The rhythm starts the movement, the movement of walking. The Pulldogs come down from the Isa Tepesi hill to get out and launch one of the lifeboats. They leave the other one there in case someone changes their mind and decides to start walking again.

Instead of heading north towards the Green Ark, they go east, towards Anatolia.

CHAPTER FORTY-ONE

Rhizomance vs Wandering

At the eastern end of the Sea of Marmara, near the city of Izmit, the Pulldogs are heading into the Kent Ormani Forest, a large park where families come to spend the day in the open air in the midst of nature. Every now and then, by the footpaths, there are picnic areas with wooden tables, stone barbecues for roasting shashlik and marinated brochettes, and roofs serving as shelters for people, small rodents, wolves, and wild boars.

The June heat forces a slow rhythm on them and frequent rests, they need two hours more rest per day than in the winter in order not to risk fevers due to overheating.

Before finding a location suitable for a nest, Alan's phone latches onto a mobile network in the middle of the forest and downloads Doctor Çakmak's answer. The email subject line reads 'Technical Report: Molecular – Biological Paternity Investigation'.

Alan removes the baby pouch with Veronika in it, hands her over to Silvia, makes an excuse, and leaves, saying, "I have to water and fertilize the plants."

"Hurry up, we still haven't found a place to make a nest."

As soon as he finds a bush, he hunkers down behind it and starts to read.

"Dear Alan, please find attached the analysis results."

On Alan's screen there are two rectangles like long ribbons full of signs and indicators. To him they look like two sheets of music, notes marked in red and black at various heights.

"I'll skip the medical jargon but, in practice, if two or more markers in the genetic profile of the presumed father and the daughter don't match, then exclusion is certain. In your case, Veronika has genetic characteristics, underlined in red, that are not in your genetic profile. *You are not Veronika's biological father.*"

Pulling up his trousers, Alan leaves the bush, stuffs his phone into his pocket and starts running. All of those seeing him streaking past jump out of the way. As soon as he reaches the front of the line, he jumps on Nicolas, attacking him from behind and knocking him to the ground. Nicolas falls forward without putting his hands out and bashes his face, then he turns instinctively, clutching at Alan to keep him at a distance, trying to stop him landing any blows. Nicolas's imposing frame enables him to hold his ground for a moment.

"Fuck you doing?"

"You think you have become so perfect, don't you? You think you can assemble and dismantle the fate of the world and all of us!"

"What are you talking about, have you gone mad?"

Alan pushes against Nicolas's legs to knock him onto his side and they end up flailing around in the dead leaves. The rest of the Pulldogs come running towards them.

Alan regains his balance, grabs Nicolas by his T-shirt and punches him so hard in the stomach that Nicolas doubles over in pain, then he kicks him in the kidneys.

At this point, when Dikran and Hakim reach them and try to pull them apart, Alan stops, unresisting. He drops to his knees and bends over, shoving his fingers into the soft earth to ground his anger.

"Why did you stay with us, eh? Why didn't you just fuck off on your own business like you always did before? Eh no! Instead you found her," he says, pointing at Silvia, her face impassive and mask-like. "And she is the highest point of your forbidden dreams."

Alan stands up again and advances on Nicolas, who shuffles backwards on all fours. Hakim and Dikran try and get between them.

"Get out of the way! This is between me and him!"

Alan shouts and shakes Dikran off. Hakim is more tenacious and attempts to act as a shield between Alan and Nicolas.

"You knew she didn't love you, but you had to try anyway, so you wanted to get to the top of the group to impress everyone. First the perfumes, then the heliotrons, then saving tribes risking extinction... and all of it because you are nothing but a frustrated ex-fatty, ex-perfumer bastard."

Suddenly Alan jumps forward to grab Hakim and throws him to one side.

Miriam grabs Silvia by the sleeve. "Make him stop! Do something!"

He lands more kicks on Nicolas's legs, chest, and head. Beginning to pant, he scrabbles for something on the ground but all he manages to pick up is a handful of grass. Alan is landing blows in a frenzy, so many that Nicolas is beginning to bleed.

"Help!" His voice comes out in a lament, like the whine of a child.

Paralyzed, Silvia says nothing, nor does she move, so Rafabel throws herself into the fight, trying to stop Alan from killing Nicolas.

"Stop it! You're going too far!"

He finally calms down and yells, "This shithead had Silvia. Look!"

He pulls out his phone and sticks it in Rafabel's hand, the email from Doctor Çakmak open on the screen. "How else do you think Veronika got heliotrons if not from him!"

She reads and her face blanches.

"Is she your child?" Rafabel asks Nicolas.

"I know nothing about this. I think you've all gone mad."

Alan presses on with his accusations. "If you think you can put yourself between Silvia and me, you are wrong. There is a chemistry between us you will never be able to break! Not even with all your nanotechnological devices."

Nicolas stays where he is, lying on the ground, aching; some of the phials in his bandolier have broken, their contents making colored puddles on the ground. Amongst the odors released into the air something attracts Rafabel's attention. She picks up a phial and holds it to her nose.

"This smell." She turns to Silvia. "It's yours.... I recognize it."

Alan comes over and sniffs it too, casting Nicolas a look of disgust.

"I knew it. I thought I smelled it at the concert in the animal refuge, but I wasn't sure."

Everyone is staring at Silvia, who is feeling totally humiliated. She wants to cry, but fights to keep control. She takes off the baby pouch and hands it and Veronika to her grandma so she can go and get the phial.

"It isn't Nicolas's fault," she says calmly. "He has nothing to do with it. Or at least not in the way you think."

She sniffs the phial, then lets it drop to the ground.

"It is my smell, I admit it. Like I admit I was with Nicolas once, the night of the Rishow." Silvia turns to Alan. "You were playing, I had been drinking. Nicolas made me an infusion to help me sober up. I stayed at his house for a few hours. What happened there should not have."

Silence. Then Silvia speaks to them all, as if the consequences of her mistake have fallen on each of them too.

"You know me, I don't tell lies, not even to protect others or myself. This time it is different, because it affects a baby. Please believe me when I tell you I didn't know who Veronika's father was, I didn't want to know because I didn't need to."

She looks her man in the eyes. "You are Veronika's father, Alan."

"How can you say that?" he hisses, turning away from her.

"Telling lies is like walking with a splinter in the sole of my foot, and this is the right time to pull it out. One day you asked me if I would have preferred Nicolas as a father for Veronika, because he's intelligent, can compose perfumes, and anything else a child could want. It's true, I care about Nicolas, we grew up together and I have to apologize to him because I have been cruel to him. In a moment of weakness I wanted his genes, I was envious of his ability to accept the transformation of his body and turn it into a strength, something it was hard for me to learn. But I never thought of him being the father of my children, and certainly not during the night of the Rishow."

Silvia tries to catch Nicolas's gaze, but he is keeping his eyes on the ground.

"Wanting something like that would be like expecting the sun to stop shining down on the rest of the world because we want it to brighten only our heads. It is true, Nicolas walks with us, but he has always been on a different path, an unknown path that every now and then – almost by chance – crosses ours. As a father for my children, I have always wanted you, Alan, even when you didn't even want to discuss having them. You climb a tree every morning to play the wake-up call for our group. You always find a place for us to make our nest for the night when the sun goes down. Your hands are dirty like mine, we have the same scars. You can be sweet, rough and melodic; you are not perfect, but you are perfect for me."

Alan turns his head a little to see Silvia out of the corner of his eye: that woman, rough and often cantankerous, has wrong-footed him again. She has turned all his suppositions upside down. He had believed she would have preferred a paternal figure like Nicolas, whereas actually she wanted the exact opposite.

"It was presumptuous of me to think I could hide all of this. I am sorry for not telling anyone before now. This is the truth."

Alan stands and takes his daughter from Miriam. Then he moves closer to Nicolas as if he were holding a weapon. "Did you hear what she said? This is my daughter."

His words wound Nicolas more than the kicks and punches did.

"I don't like the atmosphere there is around you," continues Alan. "You exert a kind of evil attraction over people. I don't want Veronika to be near you. I know from experience now."

Nicolas doesn't react but Rafabel knows the drama he is living through. She grabs Alan's T-shirt and pulls him away. "Enough! He has just found out he is a father, then that he isn't anymore in the space of a few minutes. Leave him alone. You've had your revenge."

Alan doesn't put up a fight. He takes Silvia by the hand and says, serious, "It's better if our paths separate here. Whoever wants to go with Nicolas is free to do so. We are going to continue to Göreme, then to the north, across Georgia and upwards, until we reach Russia and Ivan and Kirill's friends. Our route remains the same."

★   ★   ★

For five days, until they reach Göreme, the two groups of Pulldogs walk a few kilometers apart. The rift created by Veronika's paternity seems to be unrepairable, even though for Alan it wasn't the only problem. Nicolas's presence was an obstacle in the way of creating his idea of a 'wandering life' from the day he got it into his head to take heliotrons to tribes risking extinction. Now Rafabel and Hakim – as was to be expected, plus Shimbo and Farisa, which was more of a surprise – are not walking with them anymore, things should go more easily. Miriam is a different kettle of fish. Despite having taken the heliotrons, she has decided to stay with her family and not follow Nicolas's group. The same goes for Kenshij the taciturn.

Tasia and Pino would have liked to have gone with Rafabel, but she didn't want them to take the heliotrons, and without them the children would not be able to keep up with the group; for the moment they whine and complain three or four times a day, every time they see someone from the other group appear on the crest of a hill, or in the middle of a bare field. They complain because they can't run and join them, like any children who are prevented from doing something they suddenly have the desire to do. Then, when they too begin to see the pointy columns that rise from the Göreme valley, they forget everything else, gobsmacked.

"Look! It's like being on the moon," says Tasia.

"The moon doesn't have those things.... It looks like an alien planet," Pino corrects her.

The white clouds crossing the sky are flocks of doves wheeling between the peaks.

"Actually, they are called 'fairy chimneys'. You see those mountains over there?" Miriam stands next to the little girl and takes her hand as she points out a peak in the distance. "That's him, the Argaeus volcano created this place with lava and tuff rock. It took millions of years of wind and water erosion, of heat and cold to shape these bizarre-looking rocks."

The land around is arid, and yet under the ground there are rivers that feed springs capable of guaranteeing fertility to places where the water surfaces in plentiful springs, giving life to luxurious oases amidst pinnacles of jagged rock: enormous chimneys that look like something out of a fairy tale, strange stone mushrooms, deep valleys, and caves scattered here and there set up as shelters. A few little stone houses mark the edge of a village in the middle of brambles climbing everywhere.

"Down there, that sign points to a Turkish bath," says Alan after checking the translation on his phone.

From the side of the hill a footpath winds between blocks of stone, all that is left of an imposing Turkish bath complex, which must at some point have been used as a source of building stone.

A dribble of clean water runs into a cistern full of croaking frogs. After filling their pouches, the Pulldogs continue along the side of an abandoned farm. Its orchard of wild plum trees offers such ripe fruit that all they have to do is stretch out a hand to pick them from the branches bending under their weight.

Nearer to the village there are some families relaxing and playing *tavla*, a kind of backgammon.

As they cross the village dozens of picturesque characters gather around the Pulldogs, attracted by the arrival of strangers; dried-up old men taking their hairy, black Bactrian camels home; kids wearing Barcelona and Manchester City football shirts; tough-looking guys with tattoo-covered arms; and the inevitable women with baskets full of bits and bobs to sell balanced on their heads. The dusty air carries a smell of propane, old cars, and shashlik smoke.

The arrival of the Pulldogs livens up the atmosphere and word of mouth does the rest: in a few minutes, a handful of beggars arrive and begin to chant prayers and litanies, while yawning camels appear from behind the houses, drooling and screeching at their masters for waking them up.

An old man with a green headscarf on his head pulls out a bottle of yellow liquid from one of the saddle pockets. "Ayran. For you, good."

Alan makes a gesture to show they have no money, but the old man insists, as if the point in question is more important than money: hospitality. After drinking he passes the bottle to the others; Alan is trying to understand how far the old man's generosity will go.

"We thank you, we are looking for a place to sleep." He tilts his head to rest it on his joined hands, to explain himself better.

The old man smiles and thumps his chest with his hand. "I take you. My camel cave. On foot not far. You first outsiders after long time. War takes all things, all people."

But when Alan turns to let the other Pulldogs know he has found them a place to stop and rest, he has to resign himself to the sight of some of them – Tommaso, Pilar, Pietro, Pino, and Tasia – already lying fully clothed in the pools of water.

It does after all seem to be the village meeting place, and, judging by the signposts along the road, it must have been the largest tourist attraction in the area before the conflict with the Kurds caused it to be abandoned. The old people and the few younger people left have nothing else to do but find comfort in the waters, washing each other's hair and backs as they chat, make fun of each other, discuss life and splash around in the shadows of ancient walls and dilapidated columns.

Lying back and admiring the panorama, Tommaso, Pilar, Pino, and Tasia motion to the others to come and join them. Alan realizes he can't miss this opportunity. The last to get in the water are Dikran, Silvia, Kenshij, and little Nika in Miriam's arms.

The old man stops and waits.

Tarak's cave is ten meters deep, wide and humid. In a wider area where his fifteen camels live, the temperature is eighteen degrees, thanks to which they have slept better than over the past few days. As soon as he wakes up, Alan gets up and checks the phone hidden behind a rock at the entrance.

The first video it recorded shows Dikran coming out of the darkness of the cave to go and take a slash behind a boulder. Alan deletes it.

In the second video, Tarak drives away a couple of drunk young men come with the intention of bothering the Pulldogs' girls. He shakes his stick in the air and it is enough to discourage their boiling spirits. Alan deletes this, too.

The last video is more complicated to understand: a shadow moves slowly between the prone bodies, careful not to wake anyone. The phone is positioned on a small tripod and the movement sensor makes the video camera rotate until it focuses on the figure in the shadows. Step after step, the shadow makes its way between the sleeping people until it reaches Nika. Then it lifts the sheet, scoops up the sleeping baby, takes a few steps, and sits on a rock.

Alan shudders and immediately leans out to check Nika is still in Silvia's arms, then continues watching the video and sees, bending over the infant's face, the shadow is whispering quietly.

"Look for your smell, the rest is only a passing stink. And it doesn't matter if it is the smell of a storm or of spring, the smell of dirt or a refined essence, or that of the truth or being resigned, what is important is that it is your smell, the one you recognize as yours, the one you wouldn't swap for anyone else's and makes you get up every morning and walk. Remember, our noses point forward because it is the organ that helps us orientate ourselves."

A sound behind the shadow forces him to turn. Silvia is staring at him, her expression dark. She doesn't say a word. She is wary and cautious, as if this man has sneaked into the cave to kidnap her daughter.

"Why did you come here?" she whispers, holding him by the arm. One wrong move and she is ready to scream.

"I didn't say goodbye to her. I can't stand the idea of being separated like this...."

"Our ideas and ways of behaving are different. We have different goals."

"I still don't like it."

"Everyone has to follow their own path in life. You have to accept that."

"I know. I'm not trying to change anything."

Silvia lets go of him and lowers her eyes, as if she can't bear to look Nicolas in the eye. She had believed him capable of doing something terrible, something maybe she would have done if she had been in his place.

"What are you going to do now? Where will you five go?"

"South, I've had an idea."

"You never get tired...."

"A network of people who exchange the excess solar energy produced by their heliotrons in order to compose whatever they want."

"I bet you have already thought of a name."

"Well yes, actually, I have. My father used to deal with these marketing things, now they're my job. I'm going to call it *ergonet*. And you? Are you really going to go into the cold?"

"We're going to try, even though I have no idea what is going to happen. I mean, no one ever really knows. Sometimes I am scared to have got into something that is too difficult for me. While you are hitching, traveling on trains without a ticket, or squatting in empty houses you feel strong because you are challenging a certain kind of world, but now, when evening falls and you come to a place you don't know and it isn't your destination, with no one waiting for you, and you also know the only reason for not turning around and going home is an eccentric idea like the rhizomance...well, it gives you pause, and you begin to feel nostalgia for your roots, even if these are in a 3D-printed house on a falling-down viaduct that sooner or later is going to be taken from you to make a multistory car park or a shopping mall...."

Silvia's sigh is melancholic, then she continues, "But I hope you manage to take the nanites to the tribes and anyone else who needs them. I have the feeling that we two won't meet again anytime soon. The world, outside the cities, is so big and footpaths don't cross as easily as roads do."

"That's true, but I still believe in coincidences. Like the ones that brought us to meeting each other again after so many years thanks to the smartfume Little Simon stole."

She finds the strength to look him in the eye.

"I can tell you now. When I saw you again outside the Rendezvous, I was scared of you, someone who knew me before anyone else did. Being known too well can make us vulnerable. But now I am happy things have gone the way they have. I am happy to have helped your emancipation and transformation."

"I prefer to call it evolution."

"Whatever you prefer, the meaning doesn't change." Silvia hugs him and kisses him on the cheek.

Nicolas, filled with emotion, caresses little Veronika's forehead and gets up to go. "Can I ask you to take a photo of us together?"

She can't deny him this wish, and nods. Nicolas hands Silvia his phone.

"I'm sorry, no flash."

He moves closer to Veronika, he doesn't take her in his arms, he simply positions her. Silvia takes two photos and gives the device back to Nicolas.

Before leaving, he takes a message out of his pocket and hands it to Silvia.

"Open it, but only when I have gone."

As he walks away, Nicolas turns frequently to look back. Silvia watches him until he is a small dot in the distance.

"It has been a pleasure to walk alongside you, Nicolas," she says, turning to reenter the cave.

Alan deletes the third video and goes back. He slips down beside Silvia and notices she still has the note, held tight in her fist. After pulling it delicately out of her grip he reads it:

I was only saved after meeting you

PHASE FIVE

ETERNAL WANDERING

'In girum imus nocte et consumimur igni
We spend the night out and about and the fire consumes us.'
Palindrome by Sidonio Apollinare

'Eternal wayfarers are the days, months, and years that come and go. He who spends his life floating on a boat and he who welcomes old age holding a horse's bridle, travels day after day, and makes the journey his home. I too, I don't know since when, house in my heart the inextinguishable desire to wander, called by the wind that blows the scattered clouds.'
Basho – *The hermitage of the illusory abode*

RAFABEL COSSER
CHAPTER FORTY-TWO
Roadside Divinities

Nicolas leaves the tent and walks towards a sandy area. In the middle there is a table from which he takes a sharp pointed object, a symbolic blade Rafabel composed on the fabtotum, and moves closer to a man with his back to him. He bends forward and with a top-to-bottom movement thrusts the blade in the space between the man's spread legs and pretends to cut something, as if slicing through ropes tying his feet together.

"Our paths separate here," says Nicolas solemnly, "but everything the paths divide, the paths reunite."

Then with a light push to the back of the other man's neck, he makes him start walking. Nicolas and Rafabel watch the man until he stops a hundred meters off and starts transmitting. The packets of energy, enerbot optimized in small, compact units for extremely unstable bands typical of the desert, are invisible to anyone who is not connected to the ergonet mesh net.

Liberating anyone who so desires from the need to eat is fundamental in an environment as hostile to man as the desert. Before initiating a nanite-taking ritual, Nicolas usually says to his followers, "If nutrition bonds man to a physical space, to stop eating in a traditional way means no longer needing a territory for survival."

Nicolas goes back into his tent and Rafabel, after giving him back the fabtotum, heads towards the village.

★   ★   ★

They have composed their daily nest just outside Bir Gandouz, twenty kilometers from the border between Morocco and Mauritania. It is a flat, empty space, there are no paths or signposts, and the panorama is made up of sand, bushes, and gravel, a transit area which extends to a vanishing point dotted with signs bearing the universal sign of danger, the skull and crossbones:

DANGER. MINES – DO NOT LEAVE THE MAIN ROAD.

Coming back from the village, Rafabel surreptitiously eavesdrops on the people lining up outside the tent. The postulates have baskets and rucksacks full of objects, recyclable materials which in their innocence they think is enough to transform their requests into wishes. Some have collected small batteries and batteries from old cars, some have bundles of sticks and plastics, some have rolls of ribbons, animal skins, bones, foam pillows, shoe soles, bits and pieces of iron, portable electrical appliances, rotten fruit and vegetables, and each one is convinced they can exhume something from this cemetery of biological and technological obsolescence.

Two of them are wearing cutoff jeans and a third is wearing a blue tunic. There are faint marks on the stone chippings left by the Bedouin tents taken down a few days ago.

"It is incredible," says the first man to the second in sketchy English.

"Kind of miracle. Truly!" adds the man in the tunic, to the second man with an air of skepticism and crossed arms.

Rafabel walks over to the trio and stands behind them. She has heard hundreds of phrases like that, in the Caucasus, in India, in Southeast Asia. Wherever they have been over the past seven years, the rumors and gossip about Nicolas have gone before them, often completely wrong, sometimes right.

"They told me he transforms material like a god. How can you believe things like that?" The man in the tunic answers back.

Rafabel can't help herself; in her barely understandable English, she pushes into the conversation. "Not like a god, more like a nanite smith."

To begin with, the men are disoriented, but then, when Rafabel asks them why they are there, they recognize her and don't dare answer out loud, they just point to the man standing motionless a little way in front of them. His skin is covered in the moving arabesques typical of heliotrons. His eyes, protected by a cheap sun filter visor, are pointed up at the sky. On his shoulders he is carrying a backpack whose ribbed structure changes color every ten seconds, from blue to green to red to yellow.

"We also want to become solar plexus," says one of the men in jeans, almost whispering in a sign of deference.

The solar plexus doesn't move, he has been standing there for hours, and the only signs of life are his steady breathing and the variations in intensity of his skin's pigmentation. In reality he is uploading energy packets to the ergonet cloud for anyone requiring it in order to exchange formula or compose on a nanomat.

For the people they meet along the streets in the outskirts of the metropolises, in the out of the way villages of the savannah, or in the country towns, the plexuses are a direct emanation of Nicolas. Where he is considered a kind of god of materials, the demiurge of many myths and legends from the past, the plexus is his ambassador, a carrier of free energy with which to forge your own fate, whether this is simple survival or the start and development of a project.

Sometimes, when the requests are too numerous and needs are urgent, the plexuses stop for weeks in one place, fences are installed around them, cordoning them off to prevent the transmission of enerbots from being interrupted or impeded by some fanatic; other times they just pass through, walking slowly along routes which people approach as if they are watching the Olympic torchbearer.

"It's late today," says Rafabel. "Wait till tomorrow, after sunup, but we won't have a lot of time because we have to get ready to cross the border."

The little group nods and stays there, motionless in front of the tent's entrance, waiting for a miracle. There are another eight people behind them. Wherever they are created, the solar plexuses attract the attention of the curious, and the curious contribute to spreading the news about where the 'roadside divinities', as the people walking with Nicolas or following his philosophy have been renamed, are.

Rafabel goes to the head of the line of postulates and enters the tent, a little uncomfortable. It bothers her that Nicolas's work is often misinterpreted, that his inventions are mistaken for miracles, or supernatural undertakings, even though the formulas are out there, available to anyone as attachments and links on an open-source database. In the end, though, the aura of sacredness spinning around Nicolas scares her because the speed and ease with which these stories have spread are a sign that people still have a great need to feed their imaginations with new urban legends, even though they are shaped on the ancient ones.

"I've found the millet meal you wanted, but no vegetables. A young girl gave me two bottles of camel milk."

Rafabel leaves the objects on an oblong cellulose table. The trunk of the tree they composed the tent around has provided them with all they needed for plates and cutlery.

"That will be great, thank you."

Nicolas is finishing composing something by the bluish glow of his phone's screen. With agile fingers he is manipulating molecular bonds and every now and then adds an element from the bandolier laid out by his side. His beard hangs from his chin like a stalactite veined with silver.

He is sitting, absorbed in a hieratic pose; a dense smoke fluctuating around him smells of myrrh and flows out through the topmost opening of the tent, where it unites with the moonlight streaming in from above.

"There are still people outside. Don't overdo it today, understand? Tomorrow we have to walk thirty kilometers through no-man's land before we get to the border, and then the same again to get to the Arguin Park. Can I send them away?"

She is shocked by how greedily the listeners believe the invented stories, embellishing and exaggerating them in turn until they become urban legends, bar stories, and fairy tales to impress children and weak minds. Nobody worries about the fact they are lies with no foundation, fake news, like the things shared without thought on social media platforms. Some talk about perfumes that act as love potions, or hate potions, others talk about essences composed to cause sterility or instant healing from any illness.

People can't stop believing, and through a process of congenital narrative creation they continue to do so thanks to him. This is fine by Rafabel; when it comes down to it, fairy tales have a moral too, it is why they have lasted so long.

"Leave them, I'll talk to them tomorrow morning. As far as the border with Mauritania goes, we aren't going to take the new road, the fast asphalted one that goes to Nouakchott."

"So, which way will we go? The mine warning signposts might just be deterrents but I wouldn't count on it."

Rafabel sits down next to Nicolas, puts her arms around him and kisses him three times on the lips through his prickly gray beard. They are soft, unlike his rough, weathered skin. Rafabel looks after her skin with moisturizers and plant oils, but the same cannot be said of Nico, so much so that kids can no longer tell he is European like they used to years ago. His metamorphosis is complete. His hair is a messy, zinc-colored mane, whereas his eyes, quick but somber, possess the glint of someone who has seen the world but hasn't lost their curiosity about it.

Nicolas curls his fingers around her delicate wrist, a gesture of intimacy he repeats every night before they go to sleep.

"Hakim told me that when he first came to Italy, crossing the Sahel and Sahara deserts, the only route possible was along the beach or by following one of the guide convoys that went past during low tide. It takes more time, but now it has become a little used route with fewer eyes on it than the motorway."

"Why?"

"Because it has been mined, whereas the main road has been cleared." Nicolas lifts the object he has just finished composing and places it on top of two others exactly the same. They are round and look like big coins. "The low-frequency coil of a metal detector will ensure our safety. This evening I'll finish the other pieces."

"Can we put it off for a bit?"

"I don't know. Hakim's father is waiting for us, and the situation was already bad when we called him."

"How much food is left in the village?"

"The food has been gone for weeks. Hakim says every day they take it in turns for three people to go in search of something to eat...but there is a similar lack of food for a fifty-kilometer radius. Quite often members of the group decide not to come back."

The sound of dragging footsteps announces the arrival of the man from Mali. He is wearing a green tunic and a brown turban. His face is swollen and painful. Rafabel hurries to hand him the bottle of milk. He drinks half in one draught.

"Thanks...that's better. We can leave tomorrow night."

"Are you sure? And your molt?" Nicolas gestures to Hakim to lift his tunic.

From his ankles to his knees, large areas of his skin have come away leaving him pink, bald, and smooth underneath. The heliotrons are vibrating and making patterns under the surface of his skin, half-centimeter-wide scaly blisters.

"Let's wait for a day. It's true that if we don't walk we get ill, but it is also true that during molting we are far too vulnerable," says Rafabel.

"How are Farisa and Shimbo doing?" asks Nicolas. He's the one who supervises each walker's shedding. In seven years Nicolas has changed his skin nine times, whereas the others, like Hakim, only six times. The speed of the molt depends on absorption of the sun's rays and the intensity of direct exposure, as well as other secondary factors that influence – from one molt to the next – the time that passes between molts.

"Ecdysis. I'm afraid I have passed it to them."

"That means it'll be my turn in a couple of weeks," says Nicolas resignedly.

"I'm sorry. I hope you have a less painful shedding."

"Are you worried?"

"Shimbo says you can even die from this stuff."

Rafabel puts a handful of millet in her mouth and chews slowly. "We have to prevent them dehydrating, give them lots of water and put oils on their skin. Us three have to be in good strength to help them. That's why it would be better if you would eat, Nico. It's been thirteen days since you touched anything other than tubers and dried fruit."

"Later," he answers.

"Here, take this, Hakim, Take it to the others," Rafabel says, giving him the second bottle of milk.

"Ah…and when you talk to him," Nicolas adds, "tell your father we are on our way to Sangha. He just has to hold on a little longer."

Rafabel and Nicolas have been sleeping in the hollow of a tree trunk sculpted by the wind, and they feel more at home than they ever did on the viaduct in Rome. Above their heads a warm breeze is carrying gusts of sand.

Their bodies are intertwined, odors mixing, everything is new for her at the same time as being very familiar places in her memory to go back to every day with pleasure and tenderness. Not like the memories from her childhood, of parents from Bolzano, a melancholic adolescence, cold, scared memories, devoid of appreciation and affection. When she was sent to study with the Augustine nuns, Rafabel had a less severe and more pleasant existence than in her domestic situation. In the boarding school, because of her broad, generous smile and willingness to listen, they became fond of her even though, apart from a few subjects like physics, chemistry, and maths, they didn't manage to teach her much. As she discovered later, she was resistant to the humanities; at the most she was interested in biology, flora and fauna, but she refused any contact, or as she preferred to say, "to sully her mind" with history and philosophy.

Every now and then, out of a sense of responsibility she reckoned was purely a formality, her parents came to visit: in a nutshell their concern was limited to 'eat-study-be-good', a mantra of duties she accepted as worse than a telling off.

The only places she enjoyed being were the irrigation channels – the ancient 'rogge' – that cross the Trentino area like veins; they carried her away, above Val Venosta, along the paths of the Croda del Clivio three thousand meters high. There, in the shadow of an oak tree or a spruce, lying among grassy tufts and clumps of arnica, she would stop and listen to the sound of a stream, breathe in the smell of eidelweiss. Now though, when she hugs Nicolas and rubs against him, Rafabel could be anywhere, in Naturno or in Rome, and she would still feel she was in the right place. The nanites have made her independent of any territory, except the one she sleeps next to.

Nicolas is pretending to be asleep, comforted by the soft pressure against his buttocks.

She can smell a hint of dates and a note of ginger in his odor. Then, she realizes he is staring at her, one eye half-open: he is staring at her pale lips and the curved line of her raised hip.

Rafabel rests her head against Nicolas and sniffs deeply. She gets hints of mint, the leaves they used yesterday to clean their teeth, and from his lungs, an intimate breath encompassing paths through forests, mountains, and rocky dirt tracks.

Those smells summon images before her eyes: shadows lengthening on stones smoothed by the wind, scattered by thousands of tree trunks sticking up from the top of the rolling hills, dancing across sheets hung out to dry; colorful tents composed on desert plateaus, unrolled mats on the bare ground where they lie to rest their feet and relax, drinking three rounds of tea under the stars.

"Have you ever thought of stopping?" she whispers in his ear.

"Sometimes…but in what other life could I have saved so many people?"

"Do you miss Rome?"

What she really wants to ask is if he misses Nika, because he never talks about it and Rafabel worries that since the split in Cappadocia any reference to his daughter might hurt him. Every now and then she has been able to contact Miriam and Silvia over the Internet. Even on occasions like birthdays, she hasn't faced, or even brushed, the question with him. Whenever she mentions Rome, Rafabel gets the feeling she is referring to a long-ago event, as ancient and left behind as the Mesozoic.

Nicolas grabs an inhaler and connects it to the device. In the end, after a pause to gather his thoughts, he finds the words.

"I miss how Rome used to be, the Rome of my childhood.... I got used to living there, like you did, but I wouldn't swap that life for this. Right now our landscape corresponds to our route, our house is a path and its shape is a sinuous line drawn by desire and movement. This is how we celebrate the eternal wandering. Rome is a point on that line, a name on the map. It has historic and artistic significance, but it isn't a home. I am looking for other dimensions."

The nanomat beeps and Nicolas pours the liquid it has produced into the diffuser and puts the inhaling spout in his nose, breathes in deeply and lets out a breath smelling of aniseed.

"Y'know, Nico...first in Serra Spino, and then on the viaduct, everyone knew everyone else, and we knew we could all count on one another. We didn't choose each other, we were selected by the situations we were in. Then you came along. A man who had never walked in his life, and you transformed into this incredible man, you exploded like a supernova surprising all the Pulldogs and then, as if this wasn't enough, you turned the Pulldogs into something...else. I don't even know how to define it. Now all I can do is trust you and your visions."

He breathes in again and passes the inhaler to Rafabel.

"And then I studied you, y'know..." she continues. "When I realized what you were following, I took your decision too, and I accepted it in every way."

"Oh, so you studied me, did you?" he says, pinching her waist. "And what have you learned?"

"I've learned that sometimes accepting other people's choices doesn't mean losing, on the contrary. I've learned that a man's smell says many things about him, his identity, his ambitions. The skin's smell almost never lies. It tells if he is pure, uncontaminated, or if he has forgotten his deepest essence. And I've realized that even man is a fruit and his sweat always reveals if he is ready to be plucked or not."

Rafabel lowers her eyes. From heel to toe Nicolas's foot measures almost thirty centimeters, his stride is roughly eighty-two centimeters, over time this length has become the metaphorical unit of measurement of his thoughts. After years of movement, Nicolas has reached the point of considering himself in topographical terms: turns, forks, crossroads, zenith angles, and rolling plains. He has internalized the characteristics of the landscape to the point where walking has become, for him, a generator of daily sensations: every route provides him with new experiences, each path is not only a way of traveling from one point to the next, it is also the tool for understanding and thinking about something. It is his mind that is walking, not only his feet, and so his life has dilated, expanded, and been distributed along the way until it has itself become an element of the panorama.

However, if on the one hand Nicolas can open up a landscape like a matryoshka doll, and layer after layer see towns regress until they become farms, ruins, and barns, on the other he really doesn't know what a rugged face or a wide smile, or a sideways look, is hiding. So, even though he can climb over fences and hedges, cross plowed fields and agricultural land and transform them into ancient forest land and mountain spinneys, he cannot grasp what is hidden behind words; even though he can rewind cracked roads until they are paved with stone again, mule tracks and cart tracks, he couldn't say with any certainty why Rafabel had fallen in love with him.

For Nicolas she will always be a star, not his universe.

Not far off, the steps of the supplicants wake the nest. This morning their movements are too frenetic and anxious. In the distance a yell cuts through the tail end of the night and it comes closer, multiplied five, ten, twenty times. The Bedouins have already grasped that much-feared word, and have begun to yell and act. "Haboob! Haboob! Haboob!"

There is the noise of bare feet running on the sand, orders and incitement to hurry, the wailing of busy women and babies crying. "Haboob! Haboob! Haboob!"

Hakim appears from the tunnel connecting their tents. He looks worried.

"What're we going to do? Last time in Algeria everything got blown away. Do we stay or surf?"

Farisa and Shimbo are behind him; their faces have begun to peel and are scaly from the molt.

"We can do it, as long as we put bandages over our wounds," the woman says.

"Good, tell the others out there. Today we are not receiving, we are taking down the tents and launching the drones."

CHAPTER FORTY-THREE

Surfers on the Storm

To the south there is an expanse of dusty white ground, a wasteland of sand and pebbles sculpted by the wind into shapes like mythical animals. To the east there is complete chaos, a swelling, hundreds of meters high and dozens of kilometers across, advancing on them relentlessly. The oblong boards have been hooked to the drones which are buzzing on geostationary flight paths, while the objects that they didn't have the time to decompose have become gifts for the children.

Rafabel keeps turning to the right to check the speed of the sand monster, the 'wall of Allah', as the local people call it. It is caused by the winds from the south and announces the arrival of walls of humid air piloted by the West African monsoon. Next to her Hakim is concentrating on the swarm of drones he has composed and launched. Since they have been in Africa this big, reserved boy has started taking initiative, giving advice and warnings as if he feels an enormous responsibility now they are on the continent he came from.

"We should skirt the sea," he says to Nicolas. "It means crossing the minefield, though, and then sight surfing because that thing is blocking all signals."

"Look on the bright side, my friend, there won't be anyone on the border to check our documents, nor will we have to bribe some official to stamp the passports we don't have."

Hakim nods and pulls his bandana up over his nose, and lowers his glasses to protect himself from the vortexes and sand devils around them. Shimbo and Farisa, both swaddled in ridiculous bandages, look like frayed mummies. The drones sway above, waiting for a signal.

"What do you reckon, Harmattan?" Shimbo asks, shouting to be heard above the storm.

"It still isn't so strong, it's only blowing at sixty kilometers an hour. In thirty minutes, we will be at the border with Mauritania."

Meanwhile the other drones are de-mining the area, opening an escape route from the haboob. As soon as the metal detectors register the presence of a device in the sand, the drone grabs a stone from nearby and drops it on the mine, making it explode safely. A swarm of three devices takes less than four minutes to clear a kilometer-long path.

"Ready to *sun surf?*"

The walkers climb onto the boards and hold on to the cables linking them to the guide drones. The noses of the drones tip downwards and they start pulling the Pulldogs. The wind pushes them and the five walkers start moving, making way through the *brousse*, the African savannah, with the roaring of the storm looming threateningly behind them.

As they pass, children jump up from the safety of the leeward side of fallen-down walls or other improvised shelters, waving at them and shouting strident yu-yus. The desert surfers are an entrancing spectacle, and it feels like a duty to acknowledge the passing travelers.

Spiny wattle, dry sycamores, and karité trees dot the land whipped by the haboob. Aligned like a flock of birds, five boards cut through the burning air, speeding before the front edge of the storm, racing along trajectories, crisscrossing each other, meeting and separating again.

In the distance, by the side of the N2 road from Guerguerat to the Arguin National Park, Rafabel can see a long line of empty tents still standing. Judging by the state of the World Food Programme logo, they could date back to the 1980s. Since then the desert has advanced by dozens of kilometers, eroding agricultural areas and reducing the production of millet, the mainstay of the entire region. In the Sahel, survival is bound to a single meteorological event, the rainy season, from June to October, which they can only hope will come again this year.

"The newest tents belonged to *Médecins Sans Frontières*," Hakim shouts by her side. "I spent three weeks with them. They function as a layover for migrants coming up from equatorial Africa. There aren't any personnel anymore, aid supplies are parachuted in from above, whoever finds them uses them."

The landscape is ribbed, the colors going from yellow to red every two or three kilometers. Fine sand rises from the dunes of the same colors. If it wasn't for the zebu that appear every now and then along the dusty route also fleeing the storm, they could be on Mars.

"Then there is still a Biblical plague: locusts." Hakim reaches out and grabs Rafabel's cable with one hand. "I've seen them, I had just left Nouakchott.... When they come they block out the sky, it feels like an eclipse of the sun, but they are more scary and make more noise. They eat everything, leaving nothing in their wake."

Along the sides of the road they get glimpses of cubes of mud, each with a little window, painted white or blue. Some women are hurrying to safety, veiled from head to toe in brightly colored clothes, pink, turquoise, and yellow, over their dark skin. The running men, on the other hand, are wearing *boubous*, swirling robes with brown inserts. Some people are sitting on their heels behind a big billboard, waiting for the haboob to pass. When everything is over they will go back to the sides of the roads to find people to sell artisan digital wares to: amulets, tribal cloth, flower-patterned material, colorful slippers, leather bracelets, flasks of Mauritanian whisky, and bottles of Nestlé water.

The haboob is looming and about to reach them.

"The drones won't make it. We'll have to continue on foot," shouts Nicolas, gesticulating to make himself understood. Rafabel doesn't waste any air and gives him a thumbs-up in a sign of agreement.

"Are you ready?" Hakim shouts to Rafabel as they get closer to the low buildings of the border guards. Shimbo has also grabbed hold of Farisa's cable to help her slow down. Nicolas, further ahead, has already slowed his pace.

As soon as Rafabel can hear that the haboob is about to surround them, she closes her eyes. Then, she brakes so hard her whole body jerks forward. Rafabel is scared, not so much of the sand getting in everywhere, as being left without anything to hold on to, without any points of reference now her GPS bracelet has no signal, her deepest anxiety, which reminds her of the three worst days of her life.

When she opens her eyes again, all she can see is compact dirty yellow.

The drones – damaged by the sand – have sunk to the ground, like the boards now without a means of propulsion.

Panic overcomes her. Long moments of paralysis which fortunately snap as soon as someone, almost certainly Hakim, hands her a rope she can grab hold of. Holding tightly, she wraps it around herself once, twice, and then she tries to hand it to the fuzzy shape next to her which is probably Shimbo.

Tied to each other, they feel the first tug from Nicolas at the head of the group, and begin to run alongside the main road. Both lanes are full of cars and trucks, nose to tail, paralyzed in one direction by the haboob, and in the other, there is a queue waiting to pass the border. Each vehicle is suffocated by a reddish powder.

In this delirium of nature everything is flying around; there is a very real risk of being hit by tree branches, trunks, shards of glass, pieces of cars, and metal objects of all kinds.

The border guards on both sides, Morocco and Mauritania, are holed up in their respective huts sipping tea, when the windows begin to shake and the visibility goes down to zero in the space of a few seconds. Even though they look out of the windows to watch the 'wall of Allah', muttering a prayer, they fail to see the five dusty shadows that run past them in front of their eyes.

Over the following days, after having exchanged formulas with the traveling sellers and scraped together the materials necessary to print boards and drones, the walkers cross dried-out swamps of sky-blue salt that from a distance look like pools of water; then they cross soft dunes, descending them in a shower of pink-tinted sand. Usually, before midday, they compose their tents beneath some bushes and drink tea made with filtered seawater or from an isolated well. They rest during the hottest hours, watching the savannah changing color as the sun moves across the sky. Guided by their GPS, they move like felines during nights made amazing by the mass of stars. All of Africa's magic accompanies them as they proceed.

One worry remains, their daily battle between skin and sun. Despite all their precautions, Farisa's and Shimbo's blisters open and spread with every stage of their journey; their flesh is veined with tobacco-colored tracery, purple filaments, and rounded swellings suppurating drops as dense as ink. In a few more days the molt will be complete.

Once they are beyond the Nouadhibou peninsula, a narrow tongue of sand sticking out into the Atlantic, they begin to feel like they are entering a new country: the nocturnal realm of birds.

By the light of the moon, thousands of birds of all descriptions trace lines, circles, and irregular triangles over their heads. A covering of feathers at least a centimeter thick blankets the rocks, stratified and worn by the erosion of the sea and wind. On the water whole flocks form floating islands, others dive in search of fish or perch on the flaking carcasses of ancient jetties.

The flamingoes that are pink by day look more light blue at this time of the evening, and the solemn-seeming pelicans keep away from the noisy snipes, gulls, and cormorants.

"Ha, we should ask these guys for political asylum," Rafabel says, pointing to a sign Farisa is translating from French:

ARGUIN NATIONAL PARK
PROTECTED NATURE AREA
WORLD HERITAGE SITE FOR MIGRATORY BIRDS

"Beautiful, but we'll need more water than we can filter from the sea," Nicolas adds, upending his empty bottle.

"Right," Hakim continues, "I know a place we can get enough for a while. You see that hill down there, to the right of the blue trees? That's where I..." Hakim hesitates. He licks his dried-out lips. "...I killed Diawné, Diawné Batouly."

"The boxer you punched?" Rafabel asks.

"No, that's a different story. Diawné was a people smuggler. I met him in Bamako, five other people were with me. He promised to take us in his boat, but in the middle of the desert he threatened us with a gun. He wanted money and everything we had, otherwise no boat trip to Morocco. We argued. That's where we got free of him. With the others we walked all the way to Tangiers. If we had had nanites then...."

Hakim doesn't lower his eyes, watching the other walkers. It feels like he is reliving the day when he regained his life.

"But I remember the routes, above all the wells where we drank near the fort of Arguin."

For Rafabel, in the light of this revelation, Hakim could have invented the story about the bloke punched to death during an illegal fight out of fear of being judged and kicked out of the Pulldogs. However, he must have got the rolling gait of a boxer from somewhere.

CHAPTER FORTY-FOUR

Marabout

Two hours later, they see a group of nomads camped in the middle of an abandoned road circling the fort on the island of Arguin.

The plants that used to beautify the route have grown to three meters high, offering protection from the wind, and above all, from the sun. Here and there nanomats are dotted around giving form to a number of tent structures.

Rafabel is wearing the visor and zooms in on them. "Judging by the number of trailers, goods wagons, and engines covered with solar spray they must be a convoy...like the caravans that used to cross the desert."

The transhumance camp, followed by a herd of zebus, some goats, sheep, and gray donkeys, rests around a well and is made up of Airstream caravans, the formula for which – as far as Rafabel can see from the tag on the visor – is downloadable for free from the e-Den archive The goods wagons, on the other hand, look like they've been put together as a compositional patchwork project and African DIY. The results are a mixture of copyleft designs, open source conical cladding, pointy African turrets, plant grafts, and piebald coverings inspired by the desert fauna.

"They could be circus artistes, or a traveling zoo."

Rafabel does a panoramic of the area. The fort of Arguin is a ruin surrounded by water cisterns that have been empty for centuries.

"They are the only people around. What shall we do?" Rafabel hands the visor to Hakim.

"Do you think we can get closer without risking anything?" Nicolas asks.

"Of course, wells are the post offices of the desert," he answers without hesitation. "People go there to get and give information. They are considered free zones."

"Like hospitals," Farisa adds scornfully.

"What's more, I think," Rafabel breaks in, "anyone who looks after animals and uses nanomats can be considered a friend."

The conformation of the solar tents is similar to a carapace, and every now and then they glitter, the cold colors of zinc shining along the hard edges of the opaque shells.

"Right, I'm curious to meet the nanosmith," Nicolas says.

Having noticed their presence, two women come towards them waving their arms in greeting. Walking in their indigo robes, they look like a pair of birds flying over the sand. One of them must be about forty, her skin is shiny, not too weathered by the sun; the other looks like her mother.

"*Essalam aleikum!*" she shouts from a distance.

"*Aleikum essalam!*" the walkers call back.

Mistaking them for nomads, the woman introduces herself as Rukan. Farisa immediately asks them to speak in English as not all of them understand Arabic. The woman, after looking at Farisa and Shimbo again, begins to look worried.

"What's the matter? Are you ill?"

Farisa tries to reassure her. "It is a simple skin irritation. It is not contagious."

The older woman grabs Rukan by the arm and whispers something in her ear.

"My mother wants to know if you have the virus," says Rukan, translating the older woman's words.

"No. What virus does she mean?"

Rukan moistens her lips. "We don't know exactly. Moussa Korò says it's dangerous.... He says that in the *brousse, the dust talks* and gives you deadly spasms."

The walkers look at each other, questioningly, but also worried.

"Can I ask you who Moussa Korò is?" Nicolas asks, even more interested than before.

The older woman turns and indicates a point far off. A man, lying in the shade of a cluster of palm trees, wearing an immaculate white *boubou*, is sipping tea extremely slowly; every now and then he inhales from a diffuser connected to his nostrils, and blows a greenish cloud from his nose.

"He is Moussa Korò, our *marabout*."

Rafabel knows this word: she has heard it whispered outside mosques, and a quick search online tells her a marabout is a kind of Muslim guru, and that they have had a huge influence on the religious and social life of North Africa since the fourteenth century. They can be sedentary or wandering, following the caravans, and they act as judges, mediators in conflicts between tribes, and guarantors for the safety of the caravans. Other times, on the other hand, they are considered as political agitators against tyrannical sovereigns, or infidel invaders.

"We don't have the virus. What about you?" Rafabel asks cautiously.

Ruskan shows that she doesn't want to talk about it, but invites the walkers to go with them.

The marabout's tent is cool. Small slits along the walls of the shell accelerate the speed of the Harmattan and create a temperature five or six degrees lower than outside.

Hanging from hooks, Rafabel can see a teapot, a tray, and three cords, a pair of bags, and some *litham*, strips of cotton to wrap around the head, leaving only the eyes uncovered; everything is made of ceramics, plant fibers, and composable resin and foam, including a bandolier for the raw materials, like the one Nicolas has.

"Fate has sent you!" begins the marabout, in the usual desert greeting.

Moussa Korò is a handsome man, tall, with the proud look of someone who knows he has done things worthy of being remembered. Next to him, on his right, as is traditional, are Rukan and the old lady called Najat with her husband Taref, and next to them young Ismail and Abdallahi who are introduced as the marabout's best pupils, good boys from the best Mauritanian families. On the other side of the mat, the walkers have gathered in a semicircle.

"Can I offer you something to drink?"

"Thank you," Nicolas answers and continues, "your tents are magnificent. I have never seen a design like it amongst the desert caravans."

The marabout dexterously pours the steaming liquid from half a meter above the numerous decorated cups lined up on the low table.

"The compliments of a colleague are always welcome."

"I am sorry, but I am not a marabout." Nicolas hurries to clear this up before drinking his tea in one gulp.

Moussa Korò smooths his beard. It is so white it stands out brightly against his dark skin. Between his shaved head and the thick tongue of hair hanging from his chin, his face is calm and respectful.

"Judging by your bandolier you are also an expert *feticheur*."

"*Feticheur?*"

The marabout leans to his right and passes the loaded diffuser to Nicolas. "A molecular witch doctor, a shaman who manipulates the smallest forces of the universe to achieve great things."

Nicolas smiles while he exhales an essence of aloe and cedar. Then he passes the diffuser to the other walkers.

"It is a beautiful definition, but I prefer to consider myself a nanosmith."

"As you like," the marabout says, standing. "However, as soon as you appeared I did my research and I can assure you your accomplishments precede you, and I am sorry I can't offer you more or better, especially regarding food. Fortunately, you don't need it.... But if you would like we have some nutraceuticals, I enrich them myself – together with my pupils – with what we manage to collect and exchange in the desert."

Moussa Korò takes a tray and spreads about a dozen bars like licorice sticks as big as medallions.

"We appreciate your hospitality," Rafabel says, tasting something more elaborate than the usual leaves, berries and bushes they have been eating since they arrived in Africa.

"Unfortunately," the marabout continues, "around here climatic changes have dramatically reduced the quantity of food we are able to cultivate. Many types of plants and cereals became extinct during the last century, and the few that have survived are vanishing month by month. The nutraceuticals are easier to produce and aren't dependant on external factors...in the *brousse* it only took ten years to substitute almost all the traditional agricultural products."

Old Najat hurries to pour a second round of tea.

"So you are not a commercial caravan?"

The marabout drinks and his face becomes intense and worried. "I can be honest with you, *inch'Allah*. The caravan, as well as having a commercial goal, also has, shall we say, a humanitarian mission, but we have to be cautious because not everyone appreciates nutraceuticals around here."

Despite the surprise, Rafabel notices Nicolas is trying to subdue his enthusiasm. He looks at her with a gaze radiating happiness. Perhaps they had finally found someone who, in a different way, is following a similar route to theirs, in parallel.

"So, do you take nutraceuticals where people are dying of hunger?" Nicolas cannot hide the emotion in his voice.

The marabout nods. "Yes, also because we don't want millions of people to have to pass through industrialization and mass production. There's no need for it anymore."

"What will they do?" Nicolas asks. He has asked himself that same question, it is almost a test to understand the level of compatibility between them.

"If they don't fall into the black hole of consumerism, they can concentrate on other things...take a different evolutionary route. We will give them the opportunity not to repeat certain mistakes."

Nicolas's expression changes. "I too was so addicted and conditioned by consumerism I wouldn't even dare explore an alternative...even just the idea of doing without some of that comfort seemed like regressing to barbarism or stupidity, like the legend of the good savage."

"I'm talking about people who have nothing. They won't miss what they have never had."

"But they still dream of having something."

"You are right, perhaps it is only our Utopia...but we need it in times like these. However, we will not stop chasing it."

"We need a Utopia. Without it we would only have the present, no vision to imagine, no future to chase. We are heading to Mali, to Hakim Konerè's village," says Nicolas, pointing to his friend. "His father has called us because, as you say, the shortages are creating enormous damage there too."

Nicolas is about to reveal too much; nutraceuticals and nanites are one thing, heliotrons are a whole different story. Even though the solar plexuses have been transmitting in North Africa for some months, caution is still a virtue. When it comes down to it, Moussa Korò, despite his hospitality and good intentions, is a stranger. After letting a drop of tea fall to the ground from her cup, Rafabel takes a sip and gestures to Nicolas to wait.

A moment of silence interrupts the conversation, then young Adballahi takes advantage of the moment to pour a third round of tea.

"...so I understand your disappointment, Moussa Korò," Nicolas continues, on less treacherous ground now. "Despite everything, I continue to believe evolution is following an uninterrupted grandiose route: every generation differs a little from the previous one; when the differences add up, the species touches a genetic apex and a new creature is born. For me, every species has an unexpected origin and an equally sudden end. Evolution moves in little leaps followed by long periods of immobility. Where before it was only the meteorological climate and the natural environment operating selection of the fittest, today the engines of evolution are also the cultural climate and the technological environment."

Thirst and the excellent taste of the tea push Nicolas to pause and drink.

"Generally the leap, from one species to the next, when it happens, is nearly always rapid and clear cut. Suddenly the new arrivals no longer respond to old cues, and a return to how things were becomes impossible."

The conversation is interrupted by the shouting of a woman outside the camp.

"*Bismillah!*"[6] the marabout says sternly, motioning to Najat to deal with the situation. The woman stands and hurries to see what is going on, but a young woman runs into her at the entrance before she even gets out of the tent. The woman pushes the older Najat aside with urgency and makes for the marabout. She drops to her knees in supplication, a bundle in her hands.

"It's Zahira again, her daughter Nida has been sick since yesterday," says Najat, attempting to comfort the woman. Moussa tells her to stand, and then takes the infant in his hands.

"I know." His face bears the resigned expression of someone who has seen the same scene more than once.

Farisa begins to move towards the infant, but the mother stops her. The bandages around the stranger's face heighten her anxiety; it is only after a stern look from Moussa Korò that Zahira is convinced to let the stranger check over her daughter.

"The child's heart rate is too fast, so is her breathing, it means her blood pressure is low. Her dark yellow coloring is a sign of dehydration."

The mother only speaks Arabic, so Moussa Korò translates. "Zahira says the child had diarrhea yesterday, a high temperature but no sweating, and her skin became very dry."

"Is this the virus you don't want to talk about?" Farisa asks Moussa Korò.

He doesn't speak, as if there is some secret he doesn't want to reveal.

"Perhaps we can help the child, but we have to know what you know," insists Farisa.

"The dust speaks...the dust curses," he murmurs.

"What does that mean? This is a virus that acts like gastroenteritis, but it might spread like the flu. Does anyone else have it?"

---

6   In the name of God!

"Almost certainly the mother."

Farisa looks at Moussa Korò sideways.

"How fast is it?" Nicolas asks. He looks like he has realized the cause of the illness.

"Very, but I can't tell you exactly…" answers the marabout, but he adds nothing else.

"If it has been engineered, you and I have got to talk."

Nicolas bends forward and picks up a handful of sand from the ground and lets it run through his fingers. Farisa watches him in shock. Rafabel is just as disconcerted.

"The haboob? Are trying to say the virus might be transmitted by the dust from the storm?"

CHAPTER FORTY-FIVE

Virus

"Tell me about Zahira," Nicolas asks of his host.

Moussa inhales from the diffuser, screws up his eyes and tries to remember.

"I found her near the village of Dargol, by the Niger, about sixty kilometers from Niamey," he says seriously. "She was close to death.... I don't know how long she had been walking under the sun."

"Was she alone?"

"Yes, she couldn't speak. It took a week before she would tell us her name."

"When did you discover she was pregnant?"

"The month after, she was ashamed and mortified, but thanks to Rukan and Najat she found the courage to tell us she is a Christian. To begin with, she didn't want to talk to me, she was afraid I would send her away."

"And then?"

"Then I realized someone had abused her."

Hakim puts his hands in his hair. Rafabel's eyes fill with tears.

"Did she ever say who it was?"

"No, she felt too humiliated. Perhaps a relative, maybe a stranger, it happens all too often. I welcomed her to the caravan, we helped her give birth, the child was born healthy, then after a while, when Zahira had picked up a normal life again, well, the haboob...and the baby got sick."

Nicolas opens the Nanocad on his smartphone and puts the blood samples Farisa gives him on the display. The program analyzes the blood's composition and issues its data.

"Just as I thought, there are nanites in their organisms, I don't know how evolved though."

Separating the organic components from the nanotechnology is a complex procedure because the nanites graft themselves on the nitric bases of DNA and begin to emulate the behavior of the nucleotides, becoming, to all intents and purposes, artificial genes. You have to go deeper than the molecular level, where the importance of gravity gives way to the van der Waals forces, to untangle the mass.

Moussa Korò watches Nicolas in a kind of trance; in the same way a seed grows from the flat surface of the land to become a tree, so from the screen the 3D shape of the nanite rises, a braid of hexagons distributed to look something like a climbing plant.

"A little gardening experience from my old home's balcony can come in useful when you are least expecting it," Nicolas comments as he isolates a sequence, cuts the ends, and waits to see how the segment regenerates. The ancient technique of grafting, updated with the genetic 'cut and paste' of CRISPR, seems to work. Plants possess incredible regenerative properties, and the nanites behave the same way in recreating the specific missing structure from the original, non-differentiated one.

The 3D formula pops up on the phone and Nicolas saves it. Then he cleans the screen with his hand, gathers a handful of sand and randomly drops some grains on it.

"Now let's see if we get some kind of correspondence."

Some grains of sand stay opaque, others glitter, a sign the program has found elements of nanotechnology. The grains that are glittering possess a higher level of silica and lower level of limestone, fragments of shell and coral, including ferrous impurities, but on an atomic level every miniscule crystalline solid has undergone an alteration caused by the same agent: an ad hoc engineered nanite.

Nicolas attempts to map the recurring structures of both samples to establish any similarities. He lays one composition over the other and, relaxing his shoulders, leaves the Nanocad to do the rest.

"Here we go," he says, analyzing the results. "When I worked for the City of Rome monitoring olfactory pollution, we used *smartdust* systems, that could be activated remotely, capable of measuring levels of light, temperature, and chemical agents. In this case it really is an artificial virus. I'm sending you the formula so you can work on it with the others. The grains of sand in the haboob activated the nanites present in the child's blood."

"Incredible, but that would mean that Zahira will get sick soon, too?"

"Exactly, her immune system is stronger than the child's because the nanites have replicated more times in her organism, increasing its efficiency. Unfortunately, there is more bad news."

Nicolas closes the program, cleans the sand off the screen, and looks at Rafabel and Hakim with a resigned air. "It will be our turn soon."

"Can I help you in any way?" Moussa Korò asks.

"I don't know what will happen and I don't know exactly when. I only know that we will need water, lots of water."

During the afternoon of the next day, a coating of dust covers the landing strip of Nouakchott International Airport where the caravan has stopped. In these latitudes, airplanes taking off or landing are so few and far between they seem to come between one dream and the next.

Lying in their tent, Rafabel scratches insistently. She has an irritating sensation of pins and needles all over her body and the internal burning hasn't stopped, it's as if her blood is boiling in her veins. Nicolas keeps wanting to scratch his arms and legs, resists, and after a few seconds of respite the incessant itchiness comes back. When he lifts his T-shirt, he sees his chest is covered with large, coin-sized blisters.

"Look at this. Do you have these too, Rafabel?"

Looking each other over, they discover Nicolas's predictions were exact. Their immune systems are reacting to the virus.

"Of course, I wouldn't dream of letting you live through this suffering on your own," she answers sardonically.

Their skin is livid, with splodges that look like a bright pink pityriasis, and where the scarring is worst, it looks like a collection of unhealed wounds or scabs forming. The heliotrons, normally dynamic and agitated with shades going from green to silver, are now flaming orange.

"One thing is sure, the people who created this disease hate whoever has nanites," she adds.

"It might be worse on us; as well as gastroenteritis, these rashes reduce our solar absorption and compromise our photosynthesis."

Rafabel and Nicolas find more lumps and bumps, like burn blisters, on their necks. She goes closer to them with her nose and sniffs, jerking back immediately in disgust. There is a smell of walnut shells coming from their skin, like the smell of the stuff her mum used to use to give her a 'bit of color'. The smell evokes her memories of the torture she had to go through every time the sun shone down on Naturno. Now, like then, the smell fills her with negative feelings, it is imbued with such unpleasant memories.

In the tent next to theirs, the scene is terrifying. Nicolas had no idea the consequences of the viral infection would be so serious.

The smell of incense coming from the diffuser to purify the air has mixed with the stink of vomit. Three bodies are writhing on the beds: Hakim, Farisa, and Shimbo groan, not continuously, not together, but often enough to make their moaning unsupportable. Shimbo keeps begging, "Water, water, water," and implores Nicolas to wipe the vomit off his shoulder. Rafabel goes to Farisa, who grabs her by the arm begging to be taken to the bathroom.

"I'm sorry, I've messed myself!"

Hakim has had a relapse, in between senseless stuttering he lets out a yell in Dogon, his mother tongue.

Moussa Korò, his two pupils, and Rukan appear at the entrance of the tent. Each of them hurries to give Rafabel and Nicolas a hand. Moving through the sleeping mats, everyone tries in vain to satisfy the cries of the ill.

Moussa Korò takes Rafabel by the shoulders. "Please tell me they are not dying."

She shakes her head.

"Are you sure?"

"No! Moussa, I don't know! And I need you."

In that moment Rafabel turns and is just in time to catch Nicolas as he falls. His eyes are shiny, his mouth is swollen and his forehead is hot, but there is no sign of sweating.

He falls slowly to the ground. She follows him even as he does this.

"Wait, Rafabel! Don't faint now...." Moussa Korò helps her lie down. "Give me some water and a bar to eat, quickly!"

The pupils run and Rafabel swallows some liquid with a mouthful of the nutraceutical. Her head nods, her eyelids flutter, but she is conscious.

Moussa looks around him desperately. Shimbo has been hit the worst, the cramps and convulsions washing over him in waves every twenty seconds. Nicolas is unconscious, perhaps he has fainted, but his breathing is regular. Farisa continues to moan. Rafabel, on the other hand, neither groans nor cries. She is, though, biting her lip.

"Stop it," Moussa tells her, seeing her almost draw blood.

"To stay awake...." The shadow of a smile flickers on her lips.

"Be strong, tell me what I have to do."

"Quickly...Nouakchott...water."

From this point everything becomes confused. The sick are carried into the caravan where Zahira and her daughter are already being cared for. The camp people settle the five invalids into beds, while Tarif and Najat herd the animals together and load them into the vehicles. Then the caravan starts moving towards the town.

Rafabel catches snippets of conversations translated by Abdhallai, who is by her bedside to keep her awake. In the vehicle at the head of the convoy, Moussa and Ismail are having a heated discussion, they have left communications on speakerphone.

"...to the sea, he said so, lots of water."

"...but Rafabel said Nouakchott, she didn't mean the sea."

"...alg'habbi,[7] the sea is hot, the water is at least twenty-four degrees at this time of year. We need a cooler place."

---

7   Stupid, in Arabic.

"...a fridge. My uncle has an electrical appliances shop, but that's in Nouadhibou."

After ten seconds of silence, Abdhallai's voice fills the lead vehicle.

"A swimming pool! My cousin Azar works in a Monotel, on Route des Ambassades. They have an enormous pool."

"Right! And with a little bit of ingenuity we can lower the temperature of the water."

Ismail ends the moment of joy saying, "How are you going to convince the manager of a four-star hotel to put a group of ill people by the poolside?"

Moussa Korò presses down on the accelerator. "I'll deal with that."

Azar has the perplexed expression of someone who is finding it hard to understand. The rewards are clear: two magnificent portable nanomats, plus a course in the basics with Abdhallai, the genius of the Nanocad, to learn how to use them. However, if the manager of the hotel ever found out he would be fired and he would have nothing left but to return to Rusu, that hole of a village where there is nothing to do other than fish polluted river fish from the Senegal, and sell Arabic rubber balls no one wants anymore.

"Now, I'm going to say it again...."

Moussa Korò and his cousin Abdhallai arrived in the middle of the night with this absurd proposal and a group of people waiting in the hotel's lobby. Everybody recognizes the marabout's caravan, everybody respects his nobility, some people say he is a bit of a bandit, others say he is a cunning illegal trafficker, but he is 'of an ancient race and noteworthy class', to quote his cousin.

"...follow me, understand?" Azar nods vigorously. "I'm going to take a room. Tomorrow at dawn I'm going to go to the swimming pool and find these two lovely water insects," Moussa waves a pair of centimeter-long bugs in front of him. "The big one is a backswimmer, the small one is a Corixa."

He carries on nodding, but when he tries to speak, the marabout stops him. "No, they are not real insects, there is no danger. We are going to print several hundred to make an impression on your manager and the tourists, got it?"

Azar nods yet again, more convinced than before, and points at some emplacements in the lobby. "If you need to use PMCs we have three down there."

"We have our own, thank you.... The rest of the plan: I'll start to complain about the hygiene, threaten the hotel with negative reviews on TripAdvisor and strong criticism from here to the Sahara. Then you only have to say one thing," and Moussa holds out his hand, cuing the lines Azar has been taught.

"We should close the pool for at least two days," Azar parrots. "We just have to tell the clients we have a problem with the skimmer. I have some friends who can disinfect the area economically. They did some work for my parents in Rusu, when the river was choked with algae."

"*Ouallah!*[8] Bravissimo!" Moussa pats him on the back.

Azar is looking at Abdhallai and only manages to say, "*Labès, inch'Allah.*"[9]

He lets the group of seven sick people, including the child, pass, and settles them onto the sunbeds around the swimming pool, lying beneath two large sun umbrellas.

Moussa Korò calls him to get his attention. "In the end your manager will have a gift too. What is his name?"

"Jamaal Sow."

"Good, now go and put the no entry signs up for the pool. When Jamaal wakes up, send him to me."

Dressed in white and lying on a sunbed under a palm tree in the middle of a crowd of people sipping tea, inhaling essences, the marabout calls Jamaal Sow over with a gesture and asks him to sit down.

Rafabel is semiconscious and is floating in the pool on a lilo alongside the other sick people. As the ice cubes from Monotel's restaurant and

8   Good God!
9   It will go well, if God wills!

bar fridges melt, they are replaced. They have managed to lower the pool temperature to twenty-one degrees.

The walkers, sunglasses over their eyes and unknowingly dressed in swimsuits, are all woozy, each looked after by someone from the caravan who is pretending to be swimming alongside their lilo.

It looks like a summer party, bathers in the pool, everything very relaxed.

Jamaal suspects nothing and takes a seat by the marabout's sunbed.

"Moussa Korò, I pay homage to you. My staff and I are flattered by your presence here at our Monotel."

"Dear Jamaal, it's a pleasure to meet you, although I am sorry to have to inform you you have not paid me the same honor."

Jamaal turns white and the cup sways in his hand. With a flamboyant magician's gesture, the marabout lifts his left arm and opens his hand to show Jamaal what he is holding. His palm is right in the manager's face.

Jamaal Sow swallows his tea in one gulp and just manages to contain his concern.

"*Bismillah!*[10] It was truly polite of you to call me over to show me this disagreeable occurrence yourself," the manager says, waving Azar over.

"*Taâla ila!*[11] How is it possible these insects are in the swimming pool? This kind of thing shouldn't happen!"

The marabout slips a hand behind the back of the sunbed and pulls out a bucket covered with a cloth. "Don't get angry with poor Azar. They are things that can happen in even the best hotels. While you were sleeping, Azar and my pupils cleaned the pool. During the night it was invaded by these unpleasant guests."

The marabout pulls off the cloth and Jamaal looks shocked.

"*Elhamdou'llah!*[12] I am grateful to you, Moussa Korò. If the tourists had seen these insects...."

---

10 In the name of God!
11 Come here!
12 Thanks be to God!

"I would be very sorry if they were to post negative reviews…a dishonor for the town of Nouakchott and the whole country, a real problem considering how important these tourists are to us."

"Of course, I agree totally."

Moussa looks at Azar, who has been repeating his lines all night to be word perfect.

"We should close the pool for at least two days. We can tell the clients we have a problem with the skimmer. I have some friends who can disinfect the area economically. They did some work for my parents in Rusu, when the river was choked with algae."

The marabout pushes the bucket underneath the sunbed.

"That works for me, I know these insects. When I was young, I raised and crushed them to decorate soup instead of using prawns. My friends from the caravan are desert people, they aren't worried about these kinds of things."

Embarrassed, the manager smiles through his teeth. "But the other clients? If they see you here anyway, they will protest."

"You are a reasonable person, Jamaal, and I want to give you something. This hotel will become the pearl of Nouakchott. If you allow us to continue with our private party, I will make such an incredible covering for the pool that your clients will be queuing to take photos of it and post them on all the social networks you can think of! See those kids down there," says the marabout, pointing at his pupils working on the three holographic displays they are using to neutralize the virus. "Within half an hour they can compose a waterproof and photosensitive tent structure covered with solar spray that will last for twenty years. In addition to opening and closing as easily as a pair of curtains, it will help you save energy. What do you say?"

"*T'barek Allah!*[13] Fate has sent you! Today is my lucky day."

---

13 God be blessed!

CHAPTER FORTY-SIX

Tears

Moussa and his two pupils cheer with handshakes, pats on the back, and hugs.

"We've done it! Our first microvorous!"

Abdhallai runs to tell the others to pull the sick people out of the pool, dry them, and get them ready for treatment.

"Who should we start with?" Ismail asks.

"According to what Nicolas said," the marabout tells them, "the older nanites are stronger, and Rafabel is the only one of them who has never lost consciousness during the illness. I'd start with her. If it goes well, we can repeat the process with Nicolas, then the others, leaving Zahira and her baby till last."

Abdhallai unwraps a syringe, pulls a phial of liquid into the barrel and injects a dose of the self-replicating microvorous into Rafabel's arm.

"...it's 2,354, I know. Exactly 354 steps. Yes, I've been naughty, please, no punishment, please...."

She's raving as she has done frequently over the last twenty-four hours. Luckily, she has stopped biting her lip and scratching her face.

"...I want to play, too.... 2,354 steps, 2,512 if you take small ones, 2,185 if you run...."

Anxious to know if the microvorous is working, Moussa and his pupils gather the members of the caravan around Rafabel.

The watery solution contains a synthetic enzyme programmed to reproduce thousands of times in just a few minutes. After identifying the aggressive nanites, just as white blood cells would, swarms of microvorouses attack the infection and reduce the virus into innocuous compounds. The virus-carrying nanites are pulled in through an ingestion portal, and pushed out through an expulsion portal through which the residues of the pathogens are eliminated. "...that courtyard isn't square...it depends on your steps...before going to sleep rub your feet...yes, alcohol and candle tallow...and puff, the blisters vanish."

Four or five minutes go by, and puff, the virus has gone.

Rafabel blinks as if she is shaking off of an enchantment, but there is no magic going on here: unlike white globules which, even when they are helped by antibiotics, take weeks or months to eliminate bacteria from a person's blood, a dose of microvorouses deletes all traces of pathogens in a very short time. What is even more useful is that in addition to saving three hundred thousand bacterial sequences in its memory, the microvorous has a wireless receiver and can update the infections to intercept directly from the web every time a new virus is reported. Over the course of time, it can become an open-source dynamic immune system.

"How are you feeling?" Moussa asks her.

"Fuzzy, like being underwater."

After Rafabel, it is Nicolas's turn to receive the benefits of the microvorous, then the other walkers, Zahira, and baby Nida. After several days, the haboob and the damage it caused are finally behind them.

"Perhaps I have made a terrible mistake," says Nicolas, rubbing his shoulder vigorously. "This idea of walking without ever stopping, this eternal wandering with no stopping points, nowhere to belong to, is like a flash, a mirage, as portentous as it is unreal."

Nicolas's head is working in stops and starts. One moment he thinks everything is wonderful, the next moment he thinks everything is horrendous, as if he is hanging over a cliff and doesn't know what to do, jump or turn back.

"Alan warned me," he continues, holding his hands over his face as if ashamed, "wandering around with no destination, chasing impossible dreams of salvation for the human race, it wasn't a viable model for the long term, but I didn't listen.... For seven years I have been dragging you all around in this madness, for seven years I have made you risk your lives, in Uzbekistan crossing that damned river, the Syr Darya, and then in Kashmir and into a sandstorm as if it were nothing. If it hadn't been for Moussa...."

Rafabel strokes his forehead.

"Traveling has taught us control doesn't exist, Nicolas.... Only the illusion of control exists, and the presumption of security. Have you forgotten the days passed in the border police offices because we didn't have passports? And when they threatened us in Quetta, but they had to deal with Hakim? Then the boat sinking off Mekong, and the kids who picked us up and carried us on a swarm of mopeds all the way to Phnom Penh? If we followed you, it's because we believe in you. Each of us finds our own light in the dark, because whether we are here or somewhere else, it makes no difference."

"But if I hadn't argued with Alan, if we had stayed with the others, we would have a home now, even if only a temporary one, and places to go back to, people to see again, dates to celebrate."

He stops suddenly and Rafabel ceases stroking his hand. "You are thinking about Nika."

"Seven years. She will already have her own tastes, her own habits, likes and dislikes. The only things I know about her are what you tell me every now and then. I have seen photos and videos on the social networks, her first steps in the fields, bathing in the rivers, the smiles, the tantrums. Silvia was right, I am not her father, I never was, I would not have been capable, I can never be her father."

"You have held this black hole inside you for too long. I've never wanted to push you into talking about it because I know how painful it must be."

"You know?" The question is sincere. Rafabel knows him so well by now, and that this is the ingenuousness with which Nicolas faces life.

"Human relationships are strange bonds," she answers. "They can be there in their absence, and absent when they are present." Rafabel is looking at her bracelet, the thing which helps her not get lost, or, more accurately, to be found if she does get lost.

"Alan made this," she says, raising her arm. Then she sniffs it to smell the perfume it is impregnated with. "I have never told anyone why I asked him to make it for me. I mean, my terrible sense of direction, getting lost here and there, but that isn't the real reason."

Nicolas is listening; he is conscious of how good Rafabel is at defusing his mood swings, but this time it is different.

"Remember when I told you how my parents died?"

"Yes, in the L'Aquila earthquake in 2009."

"Right, what I didn't tell you is I was there too. They took me there even though I didn't want to go. They were convinced I would like L'Aquila, with all its medieval towns full of ancient churches and buildings, because I have a degree in architecture, but I like modern architecture. They never understood me particularly well."

One of the Monotel waiters passes with a tray of fruit which Rafabel digs in to.

"It happened on the night of April 5th, or the morning of the 6th if you prefer."

"I remember, it was really strong in Rome, too."

"When the first tremor came at eleven, I was in bed. My parents were in the next room and I asked them if maybe it would be a good idea to sleep outside. They reassured me and I believed them."

Rafabel chews on a cube of pineapple.

"Then the next tremor came at three thirty-two, the one with the deafening roar that penetrated everything, homes, bodies, memories. It was dark, and for the first time I lost my orientation in a small room. Instead of finding the door, I walked into the wall. My fingers brushed the wardrobe while I tried to find my way in the dark, complete darkness. When the floor of the B&B fell, I went with it. I passed out and was stuck under the rubble for almost eighteen hours."

"How did you get out?"

"When I came round it was dark again. I started yelling, hoping someone would hear me, but nobody answered. I thought it was going to be my last night. I had my telephone in my pocket, and after wiggling it out, I called out to my mother and father, who should have been nearby. They didn't answer. Then I thought of the only person who had ever talked to me about earthquakes, my grandfather, Bruno, who was from Udine and in 1976 survived one that registered a magnitude of six-point-four. Despite the weak phone signal, I managed to move a few centimeters and call. He had just heard about the earthquake and had tried phoning everyone. He was already there, in L'Aquila, and had been for two hours, helping the rescue services."

She stopped to eat some chunks of coconut.

"Fifteen minutes after I phoned him, my grandfather was at the address my parents *had not given him* before we left. He had found me with the MyLife365 app. When I heard his voice, I guided him; when I felt his hands, I helped him; when I saw him, I cried in the arms of that big man with the long gray beard, thick black hair, his shirt always untucked, who my mother was frequently ashamed of. He pulled me out at dawn on the day after the earthquake. We searched for my parents, and I cried again, in a different way. Smoke rose from the town and you could hear sirens all over the place. On the streets people were wandering around, strung out and unbelieving, shaking their heads like wounded animals."

Nicolas makes to hug Rafabel, but she is immobile, cold, and doesn't hug back.

"I didn't tell you this because I want sympathy. Having parents like mine was difficult. Sometimes I wished I didn't have any. At least, not like them. I know it is a terrible thing to say, but I would be lying if I said otherwise. We weren't a family, we were like strangers."

"I'm sorry. Even at fifty, some things are hard to accept."

"Especially *after* fifty years. It is not their fault if I now suffer from claustrophobia, but it is because of them that I love open spaces, if I want to lose myself, why I can't stand having a roof over my head, or walls holding me in."

Only now does Rafabel take Nicolas's hand and stroke it.

"But it is thanks to you that I no longer suffer from an 'absence of adventure' and have pushed my limits." As she says this she manages to smile. "When I was nine, I told my parents I wanted to be a vet.

They looked at me smugly and said, 'Of course, if you study and make the commitment, you can do it.' They wanted me to believe in Father Christmas, so yes, everything was possible, but not dreams, oh no. When I was eighteen, I told them I wanted to play the electric guitar in a punk group, and they said, 'Come on, be realistic, wouldn't it be better to be a lawyer, an accountant or a doctor?' Then when I was twenty-two and graduated, they took me to one side to tell me, 'At your age we were already married and planning on having a daughter,' just to add a link to the chain, and renew the cycle ad infinitum."

"That happened to me too. And I didn't manage to leave. If it hadn't been for Silvia I would still be there...."

"Exactly, so, be strong, because our home, if we can talk about homes, is all inside the fabtotum, and you are like the genie of the lamp, who can build one every day anywhere on the planet. That doesn't seem so bad to me."

"There is something I want you to see," says Abdhallai to his maestro. "We've found the author of the virus."

Moussa can't believe his eyes as he reads the signature in Arabic.

"We had to go down to one hundred picometers to find the writing, but it's everywhere, on every nanite."

جماعة[14] اهل السنن للدعوة والجهاد

"I have to go talk to Zahira."

Moussa feels a hand resting on his shoulder. "Are you sure you won't hurt her again?" Farisa asks.

"Now I know what happened, I know what they do to people, to women."

"Maybe she doesn't want to know."

"In addition to abusing her, they damaged her with nanites, like they used to do with acid. She didn't get away like I thought she had, they let her escape so she would carry on spreading pain and suffering every

---

14 Peoples group of the Sunna for religious propaganda and Jihad, Islamic terrorist group, also known as Boko Haram.

time they inject their poisonous powder into the haboob! How can I not say anything...."

"I didn't say to say nothing, I said you don't have to tell her. Because, believe me, she won't feel any better after finding this out, not for herself or for her daughter."

Moussa falls back into his chair, dejected. "You are talking as if you...."

"I've known these men since before they started using nanotechnology."

"Monsters," Moussa Korò murmurs, unable to define those barbaric acts in any other way. Horror has broken over him. Tears run down his cheeks in his sadness.

"Leave me here alone, please."

CHAPTER FORTY-SEVEN

Solar Plexus

After all this turbulence, their journey continues in serenity during the following nights. The caravan proceeds quickly – roughly three hundred kilometers each leg – along the 'road of hope', as the stretch that cuts the country diagonally from Nouakchott to Mali is known.

Seeing the length and precarious conditions of the route they have to face to get beyond the Mauritanian *brousse*, Moussa Korò has offered to take the walkers to Bamako across fourteen hundred kilometers of dusty steppe. It would take them fifteen days to walk this distance; the caravan will only take three. In this way they can recuperate the time they lost to the haboob.

The *solarized* convoy is forced to stop every now and then at roadblocks, but the marabout's reputation is such that he only needs to hand out a few nutraceutical bars – gifts for the policemen's children and wives – to be able to continue unharmed with smiles and well wishes for their journey. Other times, their infrequent nocturnal pauses are spent in out of the way villages, like the time they stop by a soft drinks seller with a PMC without printable molecules in its reservoirs. Some customers are queuing up to make basic necessities using alternative, often unacceptable materials, listening to *spin-reggae* and *blues-decalé* melodies on the African Music Network. Nearby there are the ever-present women whose clothing, beneath the icy moon, shines out in flashes of red and turquoise due to the reflective nanofibers woven into them. Sitting on stools, some are selling hand-me-down 3D-printed utensils and trinkets, and others locally made metal or cellulose crafts, like pans, buckets, cloth, and exercise books, but no food.

For almost twenty years, Moussa Korò has been crossing the length and breadth of the desert, traveling from one oasis to the next, like a family father visiting his many children and grandchildren in turn. His pilgrimages remain mysterious adventures, except for the general plan of taking nutraceutical bars to where food is becoming scarce. Then, one evening, while they are driving through the dark and chatting, Nicolas discovers his method.

"You know, Moussa, I really believe there is a sixth sense, a sense of direction, in our central nervous systems."

The soft crests of the dunes and the hard bottoms of the dips follow on from each other on the right and the left with bewitching monotony, accompanying their progress in the glimmer of the African night.

"I have seen wild animals marching across the desert. I have seen birds who look like they have their routes coded into their genes. Some people say they can work out their position by the height of the sun, the phases of the moon, and the rising and setting of the stars; and they can even adjust their direction if a tempest has put them off course."

"It's true," says Moussa from his place at the wheel. "Even if I have never seen them here, I have read that some ducks use the croaking of frogs below them to identify when they are flying over swamps."

"When we were on the Green Ark in the Mediterranean," Rafabel adds from the back seat, "the noise the migratory fish made came through the hulls. They even woke us up, do you remember Nico?"

"What about salmon who recognize the taste of their ancestral rivers?" says Shimbo from where he is sitting next to Rafabel.

"And dolphins with their echolocation which enables them to avoid underwater rocks and find safe passage through them?"

"It used to be nature that dictated times and places," says Moussa Korò. "Farmers knew whether or not harvests were enough for the families growing the crops. Stock farmers knew how many calves, lambs, and chickens were needed to feed a village in the savannah. But now, without nature?"

Moussa pulls out his phone and hands it to Nicolas. "Here, open the Nutraceutimap app."

A map of part of Africa appears on the screen, the section from the Mediterranean down to the Gulf of Guinea. It is peppered with dots; some are bright green, others dull green, others are yellow or red, and some are flashing to mark danger.

"Amazing! The bars tell you where and when to go." Nicolas turns to the man, feeling an enormous sense of empathy. "The color indicates the amount of time passed since they last ate nutraceuticals."

"My pupils designed this, the credit is mostly theirs."

"The imagination is yours though, Moussa."

He winks at him.

"I'm not kidding, I think you would make an excellent solar plexus...."

The vehicle bumps and Moussa Koròs stares at Nicolas without looking where he is driving. The long road is so straight it hardly needs his attention anyway.

"If you want to make me solar plexus, I would be honored."

Then, suddenly, there are trees. The side of the sandy road is lined with hundreds of bushes; some thick, some more slender, they look like they have only recently been planted.

"The Great Green Wall is growing well," says Hakim. It had taken him less than an hour to walk through it when he had made the trip towards Italy.

"It's not actually a dam against the desertification...but it's at least something," says Moussa Korò. "The original idea of an uninterrupted forest between Senegal and Djibouti never became concrete, and what was achieved has remained as a kind of scattered mosaic of micro projects which, as is often the case around here, start off well but end badly. We need at least ten Green Walls.... With three hundred million trees and bushes, well, then we would begin to see some results."

In the space of five minutes, in the same way they suddenly appeared, the trees are replaced by sand and stone.

The caravan enters Bamako from the northwest, along Kasse Kelta Avenue, to avoid being set upon by the hundreds of kids who can't wait to celebrate something this strange and out of the ordinary.

The streets on the outskirts are cut by a dry wind that scatters red dust over all the surfaces. All around there's a chaos of clothes and the brilliant colors of tribal cloths hanging in shop doorways, open-source packaging, junk from old archives of FreeDeeware compositions, and piles of rubbish picked over by hungry animals and people looking for materials to recompose. In the distance the bank buildings, and those of the airplane and import-export companies, soar above the rest of the city, their animated billboards and signs written in Chinese, French, and Russian. The result, in Rafabel's eyes, is that there is no longer any difference between first, second, and third worlds. The same products are available everywhere, in real time. Bejing, Berlin, Buenos Aires, everywhere, but despite this there are differences: the trends of global fashion have been adapted to local tastes, the popular songs are continuously rewritten through mash-ups, phones personalized by designers, and the same smartfumes vaporized with what is on hand nearby.

"Ah, I'm not used to all this commotion anymore," Moussa says as they go into the Hamdellaye neighborhood.

"Neither are we. It must be three years since we last set foot in a city," Rafabel says.

"Me too, but this is the best way to get to Dogon land."

"Thanks Moussa, we have made up a lot of time." Hakim leans over to see better what the capital looks like now. "Yesterday I called my father and he said he has to come to Bamako for business…. If I know him, he will have organized some transport for us to go back with him."

A wide, sandy street divides an expanse of perfectly square little brick houses printed on the local PMCs. The only difference between them is the color of the material they are composed from. This varies piece by piece, creating a prismatic effect.

"Many people came back here from all four corners of the world after the African Diaspora," Moussa continues.

Each hut has its own domestic composition device – squat and bulky like an antique cathode-tube television set – connected to the solar panels on the roof.

"In Bamako you can find Parisians from Madagascar, wandering Kenyans with one foot in Dakar and one in Cape Town, Senegalese who feel like foreigners in their own country, and above all, Afro-Brazilians from Bahia who want to go back to their motherland."

After the intersection, opposite the mosque, an area the shape of a wash bowl is home to a bus and coach station serving the whole country, and beyond this the zigzag of streets starts.

"Get ready, because we'll be at the river soon and the boarding will start."

At the sides of the caravan, a crowd of people is already building up and getting bigger at every turn. A multitude of noisy children, giggling girls, curious older onlookers, and absentminded wasters, are following the convoy, forcing it to slow down to walking speed. Some people are shouting, some are posting videos online and calling friends to join the party. The trail includes swarms of mopeds, limping bicycles, DIY hoverboards, and minicars quick to use their horns.

Before they get to the bridge over the Niger, an immense publicity banner calls the children to summer Koran school. The ad says:

MEDERSA "LA ILLAH ILL ALLAH"[15]
LASER CIRCUMCISION

Still higher up, on an installation made up of a mosaic of dozens of flat-screens, the TV is broadcasting the news of a Boko Haram attack to the north of Gao. Then, finally, in front of them is the inevitable blockade: a couple of enormous SUVs are blocking the road and three armed guards with helmets, bulletproof vests, and weapons in their hands start checking over the trailers using portable scanners they wave around near the bodywork. One of them, rather zealous, wants to search the whole

---

15 There is no other God than God.

interior of the trailer but Abdhallai and Ismail convince him to desist: the price of their privacy is high, two women's *pagne* in nanofiber capable of repelling whatever kind of dust might try to attack the material. Included in the agreed price, luckily, is an escort to the Djikoroni district where the caravan stops on the grassy banks of the Niger in a celebration of joy and general excitement.

As soon as the convoy has parked in a circle and the doors of the caravans are all about to open, the assault starts. Not for selling but for buying. The days when kids used to surround foreigners in an attempt to saddle them with something, anything, and the all-round helpers assured them they could do or get whatever they wanted, are over. There is no longer a need to beg at windows; what there is now is an immense demand for designs to share, for 3D formulas to improve, for distilled knowledge and, something even more in demand, immaterial experiences to try, like, for example, smartfumes.

The adults are standing around and mutter when something catches their eye, while groups of children and young women wander around looking for exotic materials, exploring the inside of the caravans like cats unable to curb their curiosity. Moussa's pupils hold nothing back; they show and illustrate the technical details of the convoy, and as soon as they can, they take some lucky person they have chosen randomly into a caravan to conclude an exchange of formulas.

"Let the manufacturers be many and the consumers few!" Nicolas yells, pulling out his phone and opening the Nanocad.

A swarm of people starts to buzz around him, copying and saving the 3D formulas he has set on a loop. Then he lays out the fabtotum in front of him, and turns on the pen for composing.

Witness to the explosion of joy, Rafabel sits down next to him. "Stupendous.... Look how many faces can be happy."

After connecting the pen to the phone, Nicolas connects to the caravan's nanomat by mesh network. As soon as the molecules dripping onto the nanomat's tray release the perfume of oranges and Argan oil, all the smartphones are pointed at Nicolas's formula.

"I can finally tell you what our mission is, Moussa."

Sitting and inhaling a mixture of laudanum and bay leaves, the caravan is reunited around a battery of diffusers under a tent like a futuristic Mongolian yurt. It took a few hours before the assault petered out and everyone went back to their homes, at least for tonight.

"The nanites are not a secret," Nicolas goes on, "you yourself have found a lot of information about us, but we have had to be careful, because, as you know, some kinds of technology aren't easily accepted. But if you want to, you can be part of our little Utopia."

Rafabel steps away and turns on her phone. She knows the nanite assumption ritual by heart, a sober initiation she and Nicolas have created over the years.

"These heliotrons, they are not harmful, are they?" Moussa Korò asks, a hint of anxiety in his voice.

"They are to start with. You might need to rest for the time it takes for auto-replication. The caravan will know how to help you, like it did with us. The heliotrons are actually nanites that are more complex than a microvorous and they carry out numerous functions within the organism. To be able to compose them, I started with a nanite and then invested everything I had accumulated in my previous life as a perfumer in calculation capacity."

Nicolas goes on to tell Moussa how the heliotrons could influence the evolution of humanity and outlines their impact on nutrition, on the concept of home, on the possible transformation of the economy and employment. When he has finished explaining his vision, Moussa Korò is stunned. There is an expression of amazement on his face: even though he realizes the need for the change proposed by Nicolas, he is finding it hard to believe how this will happen.

"What a scenario...even I have heard that people would be able to stop working and kids wouldn't need to study if the nanites were circulated in a free and indiscriminate manner. I also heard that the GDP would crumble together with company profits and all that shit that is economy and finance. But then I ask myself: what profit is there for a social media spammer, or someone that harvests seasonal fruit or writes online reviews for companies selling predictive algorithms?"

"If someone didn't need to find a job," says Farisa, "they would study what interests them the most, and believe me, I know what I'm talking about. Where I come from, Mokolo in Cameroon, with all the wars caused by tribal reasons, religion, and for control over resources, every woman is by necessity partly a nurse, thousands of girls share the same fate."

"And if millions of people could count on nutritional security, they could do better things than those they have to do. Maybe to begin with they would take the wrong path, but in time they would learn to find the right way," Shimbo joins in.

"You are all better than me," says Nicolas, looking at his companions. "It has taken me so long to understand this. How many environmental catastrophes do there have to be before we open our eyes? How many tearjerking presentations do we have to see about the apocalyptic fate of the world? How many more celebrity parties and international meetings will there have to be to finance the salvation of the ecosystems threatened by human behavior? How many announcements from think-tanks set up to consider sustainable economy must we read? And how many concerts full of VIPs like We Are the World, and Live Earth, must we go to? I know, these are rhetorical questions, but it's because global capitalism must be repudiated, and we are doing it, one molecule at a time. When people realize nanites are not alms given to beggars, but a tool for emancipation, it will be too late to stop the change, and when this happens humanity will have taken a step towards self-preservation. Afterwards, I don't know what will happen, but for me it is enough to have put into motion an unstoppable force in the opposite direction."

After the few seconds necessary for Moussa Korò to digest a future larger than the miniscule size of the heliotrons, Nicolas offers his wrists to the marabout.

"And now I am asking you, do you want to become a solar plexus on the ergonet network?"

Moussa Korò nods and Rafabel reappears holding a diffuser and a phial containing an emerald-green liquid. There are two inhalers coming from the diffuser. Nicolas and Moussa insert them into their noses.

"May heliotrons free you from the 'demon of the place' that imprisons you in a territory, that makes you the slave of its habits, that takes away your curiosity to discover the world, that takes possession of your adventures and swaps them with imitations of life."

Nicolas inhales and his chest swells like the trunk of a baobab full of water to survive the drought of the desert.

"May the nanites free you from the 'devil of the place' that prevents you from realizing your infinite identities, restricting you to only two: citizen and consumer."

Nicolas pushes a cloud of smoke from his lungs into Moussa's inhaler and Moussa inhales hard to receive the heliotrons. Then Hakim invites Moussa to stand, and asks him to turn so his back is to Nicolas.

"May the nanites bring you the breath of life from the wind and enable you to walk unharmed in the heat; head towards the dawn in front of you and chase it so that you never have to see it go down."

Nicolas is holding the pointed object Rafabel gave him. He grips it, moves closer to Moussa, and thrusts the blade into the space between his spread legs. With a symbolic gesture, Nico cuts the ropes binding Moussa's feet.

"Our paths separate here, but what paths separate, paths bring back together again."

An inebriating perfume accompanies Moussa Korò's steps after Nicolas encourages him to walk by pushing him gently on the back of his neck. It is the smell of hibiscus from the hot drink prepared by Najat and served to the members of the caravan in many little cups by the old man Taref.

CHAPTER FORTY-EIGHT

Griaule

The space where the caravan had been camping is empty. The absence of Moussa and his wandering tribe makes them sad, though he left the walkers with words of hope.

"At over seventy years old you have given me an excellent reason to continue crisscrossing the desert and to not stop doing what I have always done. Since I left Marseilles at least. Ah, but that is another story!"

Then the convoy left, Rukan sang a song of farewell, and the others waved their hands. A crowd of people followed the convoy right out of Bamako.

Now Hakim and Rafabel are walking on faded ground, a strip of land coasting the banks of the river with tufts of grass, bushes, and a muddy marsh. Every now and then they walk on planks of wood that make the going easier. In the middle of the river dozens of pirogues row by loaded with people. Some are carrying animals, sheep, goats, and chickens alongside mopeds and parcels of various sizes. It's easy to see the tourists, they are on *pinacce* with transparent sheeting protecting them from the insects.

"The Niger is only just navigable," Hakim says with a sigh.

"This is, in theory, the rainy season," Rafabel replies, brushing a cloud of red dust from her clothes.

There are hundreds of weavers, one next to the other along the southern bank of the river, using old 3D printers to make carpets, colorful *pagne*, and tribal patterned cloth that they display on the ground like murals, as though the embroiderers were curators in a modern art gallery.

When they turn back, Shimbo is busy decomposing the hard parts of the camp and the other walkers are sitting under the trees.

"Be patient. My father is not really the punctual type," says Hakim, apologizing.

A handful of little girls have surrounded Nicolas and are watching him composing a pair of shoes, except for one who comes forward and daringly tries to touch the hologram and the filament coming out of the nozzle to find out the secret. Hakim is wrong, though, because not even a minute later the tall figure they see leading a dozen kids towards them turns out to be Alphonse Koneré.

"Hakim!" he exclaims, pulling his son into a hard hug. "How many years since I last saw you?"

"Twenty-four, a whole generation. Hello, Dad."

Alphonse is wearing his best tunic, an elegant indigo *boubou* and a Dogon beret made out of the same material. Four ritual scars cross his face from temples to eyebrows. His hair is plaited tight to his scalp and his skin seems to reflect the blue of his clothing.

"Yup, a generation...." And as he is speaking, Alphonse rests his hand on the head of the kid standing next to him. "This is your brother, Askalu."

"I don't believe it! I've seen your photos on social media. How old are you? Ten?"

"Eleven, just had my birthday."

Slim, wearing a too-big T-shirt, Askalu takes off his sunglasses and smiles, his barely visible moustache smiling too. His skin is the color of chocolate and his hair is done in peppercorns.

"Well then, let me shake you by the hand, like adults."

His cheerful eyes light up as he shakes Hakim's hand proudly.

Hakim turns and introduces him to the rest of the group. "This is my father, Alphonse Koneré, one of the wise men of the village of Sangha, and these are, well...the walkers."

"Hakim has told me so much about you. I hope you'll be able to help us..." Alphonse becomes serious, "...because we could really do with some." Then he takes his son to one side and starts talking in Dogon.

After a minute, Hakim translates for the others.

"A friend of my father who moved to Bamako works for Bani, a bus company running routes out of the *Gare Routiere* in Songoninko. He says this evening there's a bus going to the markets on the plateau and can give us a lift."

Nicolas slips on the shoes he has just composed and they all start walking, with Askalu ending up on Hakim's shoulders.

"The first word you have to learn in Bambara language is *tubabu*," says Askalu. Farisa and Rafabel are sitting next to him. Further up the bus Nicolas and Shimbo are hunched down; they have no option but to share their seat with a lady wearing a flowery dress who is smiling generously around her.

"Why?"

"It means white, like you are."

"Don't the Dogon have their own language?"

"Yes, but everyone knows Bambara."

Behind them, Alphonse and Hakim are discussing something. The back rows are a raging chaos of people and tourists all tightly packed together. The last to arrive squash together holding their backpacks, bags, and suitcases in their hands. Human cubs and animal cubs share their relatives' laps and patches of burning-hot floor.

The bus has only just started when Rafabel realizes, for the first time, in the middle of Africa and looking at faces and bodies all squashed together, her 'lack of color' beneath the epidermis darkened and covered by heliotrons. She, unlike Nicolas, has always been pale. She is looking at her arms now and Askalu smirks, laughing at her a little because her sparse blond hairs are more visible than his. The young boy gets up his courage to ask a question.

"What have you all done to your skin? Hakim has it like that too."

"It is something Nicolas invented and then passed on to us."

"It's strange," says Askalu, brushing his fingers over Rafabel's forearm. "Does it hurt when it moves?"

"Not at all. This modified skin helps us walk for days and days without getting too tired."

"Really? Then I want it too! Every morning I have to walk two kilometers under the sun to go and fetch the water."

"They send a young boy to fetch water?"

"Of course, it is no longer a dishonor."

Rafabel punches Askalu gently and pulls her sleeve back down; some of the other passengers are peering at what she would rather they didn't see.

"Don't tell anyone else. There are people who would not appreciate this skin."

"OK. Where I come from we have saying. It says, 'Wisdom is like a skin bag, everyone has their own.'"

Looking through the windows as they are leaving Bamako, they can see a series of small houses, market stalls, and nonspecific secondhand sellers, commercial signs, and sellers of juices and drinks in cans. In a trance induced by the soporific effects of seeing so many objects per square meter, Rafabel almost doesn't notice how heated the discussion behind her is getting.

"What are they saying back there?" she asks Askalu, telling him to whisper a translation for her.

"I didn't get everything. Alphonse said he had no choice, that a Chinese company promised him a system for...how do you say...for bringing water."

"An aqueduct?"

"Yes, they promised him an aqueduct instead of the old dams built by the NGOs."

"What did Hakim say?"

"He said Alphonse has made a mistake, and it's the fault of Griaule and onions."

"Griaule? The French ethnologist? He must have been dead for a hundred years. What have onions got to do with anything?"

"I don't know. Then Alphonse got angry and said Hakim left Sangha, so he has no right to criticize. Then Hakim got angry and said he left because Alphonse never listened to him, like now."

Suddenly there is a pause in the argument. Rafabel peeks behind her and sees the father and son, offended, back-to-back.

"How did it end?" she whispers to Askalu.

"Hakim asked what the Chinese company wants in return. Alphonse answered 'the cliff'."

At dawn the next day, after crossing the Niger and coming level with Mopti, the bus passes an old ESSO petrol station, burned black. For kilometers and kilometers, the trees have been burned. The temperatures in the area have probably led to the spontaneous combustion of all the plants and trees across the blackened ground.

The asphalted road climbs up along the backbone of the plateau, then for another ten kilometers snakes like a burning ribbon through hell to the village of Bandiagara. From there on it becomes a dusty track. A solitary road sign says:

DOGON PLATEAU

A little after dawn they have to stop to change a flat tire.

Everybody gets off the bus. Nearby there are four little houses, a village of sorts, called Dourou. After ambling two hundred meters, the passengers find some shade in the yard of one of the houses where there is a straw canopy. Rafabel takes the opportunity to move closer to Hakim.

"Can I ask you why you were arguing with Alphonse? I don't want to interfere in your affairs, but it is important to understand how our proposal will be received."

"It's complicated...."

"I thought it would be, but try. We have to find the best way to put forward our solution."

Hakim turns to check where his father is. Alphonse is chatting with two western girls One is wearing safari trousers and a T-shirt that were distressed even before their excursion to the escarpment, the other is wearing a post-apocalyptic frayed top and trousers. They both sport A/R visors over their eyes with which they are transmitting their vlogs live from Mali.

As they start walking back, Rafabel slips her arm around Hakim's shoulders, encouraging him to continue.

"We used to grow millet and then Griaule brought onions. We had the rains, and when the rain got less Griaule had dams built so we could grow more onions. We used to have tribal rituals that had been ours for thousands of years. Griaule and his troupe of anthropologists started to take everything, and now we have a ballet company made up of professional actors and dancers. Now half of Africa no longer sees us as the Dogon, but as onion sellers and dancers!"

"What does your father think about that?"

Hakim shoots a dirty look behind him. Alphonse is gesticulating, trying to please the tourists.

"Wait till you get to Sangha, you'll see. I'm afraid that what you'll find won't be the escarpment where I was born and raised. Even then it was not the same as the cliff where my father, my grandfather, and my great-grandfather were raised, before Griaule came."

The driver lets out a yell, waving his arms around to let them know the bus is ready to leave.

"Why is the escarpment so important?" Rafabel pretends she doesn't know what Askalu told her, to find out if the kid missed anything.

"My father came to get us at Bamako because first he went to talk to a company that wants to invest in mass tourism, here on the cliff, and guess what part the Dogon have to play in this project? The resort hosts...." He shakes his head in disappointment.

"Perhaps he was desperate. Try and see it from his point of view."

Hakim gets agitated. "You too? I get his point of view! I know my father, he wasn't desperate. That's why we were arguing, because he says the Dogon have to progress, that tourism will help finance the survival of our culture."

"So perhaps he will like our solution. We will be giving them a ready-to-use evolutionary leap."

As they walk back to the bus, Hakim seems disappointed.

"You don't know him though. He saw the advertising on the Internet and expects free nutraceuticals falling from the trees and growing in the ground like onions. Plentiful food that never stops multiplying on its own. That is the kind of progress he brought us here for."

Rafabel remembers something: "One day, I must have been fifteen, my grandfather took me with him to the top of the Croda del Clivio where a friend of his lived, a hermit. When we got there my grandfather asked him, 'I'm always hearing people say, "It's progress, old man!" And I never know what they mean. Perhaps you know, what is progress for?' The hermit lit his pipe and thought for ten minutes. I was getting bored, but my grandfather sat in silence next to him. Finally the old man nodded and gave him an answer. 'It keeps people busy.' My grandfather wasn't satisfied with the answer, and he said, 'Yes, but why?' Then his friend smiled and this time he didn't have to think about it. 'It makes them feel important.'"

"We should remember that when we talk about the nanites to my father."

"Is there anyone else in Sangha who could help us convince him?"

"Maybe, before talking to Alphonse we could go to him."

"If it doesn't work," says Rafabel, "we can always dress Nicolas up as Griaule...."

"That's not a bad idea."

Hakim manages to smile and when he gets back on the bus, Alphonse is no longer in his place. Shimbo is playing with Askalu, and Rafabel goes to lie over Nicolas's legs. He is in the same place, not having left the bus.

Hakim's father has gone to the back row with the western tourists, so Hakim goes to sit next to Farisa who didn't leave the bus either. She leans her head on Hakim's shoulder.

CHAPTER FORTY-NINE

The Cliff

They are back on red ground. The rickety trees are all the same height. Dozens of huts clustered together forming the outpost of a village appear every now and then amidst the rocks. They can see traces of spontaneous combustion, a mixture of black earth and carbonized embers, and here the bus stops.

"From here you have to clamber up to Sangha," the driver tells them, waving vaguely to the left. The pebbly slope reaches right to the base of the cliff, then it becomes an escarpment. "I have to get back to Bamako."

As soon as they get off the bus, all the passengers, except for the Dogons who head off to their various home villages, stand stunned. The cliff face is like one of those walks of slate full of fossils where you can trace the whole history of an area backwards through the geological eras. The footpath is a scar of compressed earth crossing a landscape dominated by plant and mineral worlds indifferent to the presence of man.

"Welcome to the Natural and Cultural Sanctuary of Bandiagara, UNESCO World Heritage Site! My home!" exclaims Alphonse in a loud voice, just like a tourist guide.

Hakim whispers in Rafabel's ear. "Now you have to hear the official presentation."

"The Dogon people were born in an ancient and mysterious land. A land where the wind and rain, commanded by the gods, eroded the cliffs of the mountains with their strong invisible teeth, creating a flat, monotonous highland like the plateau."

Alphonse motions them to follow him.

"He always starts with these Amadou Hampâté Bâ quotes," says Hakim quietly so his father can't hear him. "When I was a kid, he used to take me with him...he wanted me to become a petite guide like so many other Sangha kids."

"To reach Sangha we have to climb that laterite fortress," Alphonse continues. To do this the group moves forward along the mule track, crossing longitudinal fault lines and using their hands to help clamber up the more difficult parts.

Hakim and his father coordinate the group's movements, helping with the 'Dogon ladders', forked rods with carved grooves to create steps, in the more difficult places. The only one having fun is Askalu, who is tripping along through stones and boulders as if he knows them all personally. He climbs quickly, following a trail that millions of feet have pressed into the sand before him. When he looks behind him, he pulls out a tooth-brushing stick and starts rubbing his teeth.

When they are about halfway up Farisa has a panic attack. The sheer cliff looms threateningly one step ahead of her. Shimbo looks at her worriedly and calls for help.

"Can someone come and give us a hand?"

Hakim climbs down like a spider, reaches out and holds on to her. She clings to him, closes her eyes and only opens them again ten meters higher up where the path becomes less difficult.

On top of the plateau, a desert of rocks extends as far as the eye can see. A lunar landscape crossed by strips of yellowish savannah. Above this, mirroring the ground below, big, dark, scary clouds hang in the sky.

"Three months of torrid heat and here comes the rain. It always likes to make a dramatic entry," Alphonse says.

Soon afterwards, the first rain thunders down, brought by gusts of warm wind. It only lasts for five minutes. The conditions of the path are getting worse and worse, muddy and full of smooth purplish stones. An aerial for mobile phones shows them Sangha is close by. The corrugated outline of the plateau is rudely interrupted to the south where it plunges three hundred meters to the Seno-Gondo plain. Walking down the drop, they reach the edge of the ravine.

"Isn't it incredible?"

The plain is blanketed with a grayish patina and the crests of dunes poke through in the distance. The constant dull noise coming from the left is a stream throwing itself over the cliff, forming a waterfall of nearly two hundred meters.

Hakim grabs Farisa's arm and motions to the others to follow him.

"C'mon, I want to show you where I was born."

They all lean out from the edge of the plain; below them they can see the straw roofs of Banani, a tiny village at the base of the escarpment.

"My house is down there, the one with the green roof. When I was four, we moved up to Sangha. My father can't stand insects and he says higher up the air is healthier."

There's a lot of fuss as they enter the village.

"Tubabu! Tubabu!" There are kids shouting, laughing and pointing at those strange people who don't have curly hair. The cockiest get closer, grab Rafabel's hand, comparing it to their own small and dark ones. They are bewitched by her heliotrons. They know about tattoos, but they have never seen them move. The kids stare at Nicolas, intimidated. His gray beard is the same as the fake one on a tribal mask. Then there is the man with the elongated eyes, Shimbo; he makes a big impression on them...he looks like one of those people who come in the groups of tourists and buy everything.

Hakim hugs everybody, shakes hands, hits high fives, and exchanges shoulder bumps of recognition; his eyes are wet with tears, and after a few steps his emotions overcome him and he drops to his knees. "I'm back."

Farisa helps him back to his feet. "Up you get, c'mon, take us on this tour, will you?"

He walks in front of them, after twenty-five years he needs to get his bearings. It doesn't look like the houses have changed; they still blend in with the rocks. They are circular with semi-spherical domes, a nonintrusive architecture reminiscent of earth-and-traw igloos. The narrow lanes create tongues of shadow hugging the walls. There are no longer children taking livestock to and from the village; instead there are pipes of a composed material and every roof has a coat of solar paint feeding photovoltaic panels, and rudimentary PMCs.

The clay terraces and granaries are thrown together in a huddle, exactly like when he left. He brushes his hand over the walls and carved doors. The buildings are damaged, crumbled in the sun, badly looked after. Suddenly, a number of drones fly past over their heads.

"What are those for?" Hakim asks his father.

"They are Chinese," Askalu answers for him from behind. "They are called travelers' drones. A client can visit the village by piloting a drone from the comfort of their armchair at home. If they pay a bit more they can even go inside the traditional houses, talk to a guide or watch when we do dances."

Just then a group of young people appear leading a flock of drones around as if they were a pack of dogs. Others are selling tourists local routes and ritual experiences to participate in through their visors.

"This is what I was talking about on the bus," mutters Alphonse. "We are tortured by the drought and instead of leaving the young people to follow the mirage of modernity somewhere else, why not introduce a little progress here to Sangha, without leaving the old people in the indifferent hands of fate and caprice of the climate?"

Hakim feels called upon to answer, but he doesn't get the chance because Alphonse turns a corner and raises his voice. "Talking about old people...."

A tall guy with a shaved head and a stubby cigar in his mouth appears in front of them. He is wearing a pair of ivory A/R glasses and an orange T-shirt bearing the words KEEP CALM AND CUT IT SHORT in black. He is holding a glittering razor in his hand. Inside the hut where he works, a man in a chair with helpfully welded armrests is looking at himself in the mirror.

"Do you remember Uncle Blaise?"

Hakim runs to embrace the man with the razor. Even the client doesn't care one bit about his haircut and also jumps up to greet his old friend.

"Let all who don't love you be damned!" Blaise mutters.

"Let me be damned if I don't love you!" Hakim echoes his words.

There follow more family reunions and formal introductions for the Pulldogs.

"This is Blaise Konerè, village medic and barber."

The shop's furnishings are minimal; a mirror like a porthole, two towels decorated with the Eiffel tower, and a shelf made out of a tree branch where the tools of his trade are hanging: two pairs of scissors and a pair of solar-powered clippers.

Pages ripped out of fashion magazines have been stuck on the walls, so when clients don't know what they want they can choose their style from these. There are also posters from Roots Travel for tourists looking for something exotic: two identical models dressed as voodoo priestesses, kids challenging each other onboard colorful pirogues, a chubby baobab pierced with climbing grapnels.

"What style are you doing, uncle?" Askalu asks.

Blaise takes off his glasses and hands them to his nephew. "Look here, tell me how I'm doing."

The glasses have an app which projects hundreds of styles on a client's head. They can be changed by the blink of an eye.

"It's terrible!"

"Hey! How dare you!" the client shouts, while jokingly trying to grab him. He sits again even though his barber is well and truly distracted by now.

After exchanging a few jokes in Dogon with his uncle, Hakim turns to the others. "Blaise says we can stay in his house as long as we want."

"C'mon then, *work can wait*," his uncle adds, sending his client away and closing the hut with a padlock.

The group walk down lanes of light and shade where there are habitable truncated pyramids, prisms, cubes, and cylinders. It is all ruined and sun-blistered though, the clay and straw walls ridged like the skin of an elephant.

Blaise's home is a courtyard created by round huts with conical roofs. The facade is made up of small cells, the ground floor pierced by a low doorway. The front has ten little niches for swifts to nest, and flat stones decorate the corners.

The Pulldogs settle down on the floor. Blaise's wife and daughter appear immediately. The woman is carrying a vase of water on her head and her daughter has a tray of millet couscous with a vegetable sauce.

"The food is not necessary, the children need it more than us," Hakim says a little too brusquely. Blaise frowns; Hakim's reply has left him wrong footed.

"We'll try some anyway, we would not like to offend you," Rafabel hurries to say. "The water is extremely welcome."

Nicolas drinks two bowls and after quenching his thirst asks for everyone's attention.

"I want to thank you in the name of my friends for your hospitality and generosity, Blaise. You know why Alphonse asked us to come here, but we still don't know how you want us to help you. Consider us as emissaries from a different civilization, less advantaged in economic terms than the western one, but more evolved...."

In the evening, as the Pulldogs begin to get ready to go to sleep, they pass the diffuser around discussing their next moves.

"My uncle said he would call a meeting of the elders in the *Toguna*, the Council House where all the important decisions are made. Alphonse is one of the wise men and our proposal will be considered, discussed, and voted on by all the members. We will have an answer tomorrow."

"How likely is it that they accept it?" Nicolas asks, pouring opoponax oil in the diffuser. The balsamic essence is pungent, made from a bush of myrrh a kid in Bamako gave him. The kid's mother was Ethiopian and used it to soothe nerves.

"I have no idea. My father will have his supporters, people who he has convinced of the need for nutraceuticals. Blaise, too, he appreciated your demonstration about how heliotrons could make the escarpment self-sufficient, independent of foreigners and the climate."

"Perhaps it would be better if some of us could talk to them about our proposal in person," suggests Rafabel. "Heliotrons can scare people, or be misunderstood by those who know nothing about them."

"This is impossible, only the wise men are allowed to take part in the council, and for the Dogon to consider you wise you have to be over sixty."

A sound from behind makes them all turn around. Somebody is trying to climb over the courtyard wall.

"Who goes there?"

Two pairs of arms appear, hugging the top of the clay wall, soon followed by two heads, both wrapped in orange scarves.

"What a good smell…can we sniff too?" They hear the question from outside.

Hakim goes to open the door and invites two women, identical in their beauty, to sit down. Their eyes are like big black wells, their full lips thick and dark, and their cheekbones high and elegant. They are wearing necklaces that accentuate the slenderness of their necks and their earrings are made so they weave in with the spiral of their ears, down to the base of the ear and around the lobe to end in an elaborate pendant.

"Do we know each other? What family do you belong to?"

"Diallò," answer the girls together.

Hakim looks surprised at the name. "How old are you?"

"Twenty-six."

"I can't believe it. Are you the famous Diallò twins? When I left you had just been born."

"Yes," the twin without nose piercings continues, "I am Nyala, that's Aissa."

Hakim offers the diffuser to the twins. After sniffing the aroma that has spread through the area, they can finally inhale it, deep into their lungs, like a breeze tickling their nostrils.

"How is your mother?"

The twins are inebriated, almost dazed by so much olfactory deliciousness. They blink their eyes and it takes them a few seconds to get a hold of themselves again.

"She's well, thank you. It's hard for her, like for everyone else, but she's well," Aissa answers.

"Why are you famous?" Shimbo asks, without taking his eyes off the girls.

Hakim takes the diffuser back and passes it to Farisa. "To understand why they are so special, I'll have to tell you the genesis of the world according to the Dogon people."

They all look at each other and smile.

"That's exactly what we would like to hear," Shimbo says, even more curious now.

To embellish the story, Hakim stands up, grabs a stick and uses it to draw an elliptical shape on the ground. "From the egg that existed before everything, the god Amma created the sun and the moon using balls of clay wrapped in eight spirals of gold and silver."

Inside the primordial egg, Hakim traces symbols representing the various elements. "Then he made Tenga, the Earth, with the body of a woman, and he wanted to take her violently but an ant nest prevented the union. So the god, infuriated, destroyed the ant nest and inseminated the Earth with his seed, the rain."

The elongated ant nest looks like a uterus. Then Hakim rubs out the upper part with a few flicks of the stick, imitating infibulation. "From this violent union Yourougou the jackal was born and Amma put an end to cosmic chaos by transforming the jackal into a fox. Then the water, divine seed, penetrated the Earth's womb and so Nommo was born, the first rough attempt at a pair of twin amphibians."

His drawing of Nommo is half fish and half human, with stylized breasts, fins and arms just hinted at, and a head with a number of antennas.

"Nommo, the god of water, returned to the sky and from on high saw mother Earth lying naked and unable to speak. So he wove together fibers from plants already created in the celestial regions, and dressed her. Adorned in this way the Earth gained a language: fiber was the first known word, a good word. Yourougou the jackal, the first son, born of violence, wanted to own the word and put his hands on his mother's fibers, it was an incestuous gesture with serious consequences: the jackal, having gained the word, revealed to the soothsayers the plan of God and made the Earth impure in his eyes."

Hakim moves the dirt around, making all the drawings unclear.

"Amma put an end to the disorder by sacrificing Nommo, the male twin, and with his dismembered body parts purified the universe. Then he exploded the seed of the fonio containing all the elements of creation and placed an ark alongside the resuscitated Nommo, and his 'babies', four pairs of twins, all the same and each provided with male and female souls, these are the ancestors of all humanity."

Hakim draws a boat in the sand, the prow pointing downwards.

"The ark came down from the sky and crashed into Yourougou's land, northeast of Bandiagara; the ancestors generated eight descendants, each reproduced on its own because they were both male and female. Nommo, however, realized human life wasn't adapted to double beings, so with circumcision removed man's feminine parts and with excision removed woman's masculine parts, restabilizing order."

Hakim drops the stick and sits down.

"This is why the woman who gives birth to twins is considered a kind of divinity."

"Then this is the right perfume for you," says Nicolas, passing them a diffuser he has just loaded with a different aromatic composition made from mastic resin and juniper.

The girls inhale voluptuously. "Here in the village we've never smelled anything like this."

"I hope it is worthy of your rank," Nicolas says.

"Our mother would love to get a sniff of this perfume," Nyala says timidly.

"Run, go and call her," says Rafabel, annoyed at the attention the men are giving the new arrivals. "We have to go to bed soon anyway."

Nyala vanishes and in the space of two minutes she comes back with three ladies: the twins' mother and two other women who couldn't resist the desire to sniff something perfumed. As soon as they have inhaled they look at each other, full of wonder, as if they have been dragged into a place that is not the escarpment, a cool, inebriating place. Immediately afterwards the oldest of the three puts forward a question.

"May I spray some on our *pagnes* so I can smell it tomorrow morning too?"

"Yes," says one of the two women, "can we bring you some clothes for you to perfume?"

"Best not," says Nicolas. "Have you ever used a 3D printer?"

Their answer is a long silence.

"Would you like to learn to use one?"

The twins nod in unison.

"Come closer, and I'll explain how it works in general terms. Then tomorrow or another day you can come back and I'll show you in detail."

They all go to bed, except Shimbo, who stays there, unable to pull himself away from the magnetism of the twins, whereas the other women go home too.

Nicolas turns his smartphone on, opens the Nanocad and selects a simple molecule: hydrogen peroxide.

"You see, you can change the molecule by changing the atoms it is composed of. The program tells you if it is stable and then suggests materials you can get the necessary elements from to make it. If you keep the phone's camera on while you walk, the scanner analyzes what you see and will show you useful elements nearby. With GPS it works over larger distances because it makes use of Big Data. When all the materials are ready, the Nanocad tells you how much energy you will need to assemble the product and how long it will take. Then there are an infinite number of additional plug-ins which make everything easier or more complex depending on what you prefer. The system is dynamic, in the sense that if you give it more energy it adapts and speeds up the process, or it slows down if you don't have the power. Do you know Lego bricks?"

"Yes, we've seen them in the market in Sevarè," the twins answer together.

"So, this more or less works in the same way. You can learn how to use it within a few weeks."

"But the phone, the connection, the program...which do you use?" Nyala asks.

"Excellent questions. You can compose the phone in blocks, the formulas for the pieces are available through the e-Den portal, together with an open-source Nanocad program. As far as the connection goes, you just need to use our solar plexuses' mesh net. We call it the ergonet."

When they smile happily, Shimbo experiences a pull of joy, and turns away to try to sleep. Then their smiles transform into a more serious expression, almost fearful.

"What did you do there?" Aissa asks Nicolas, pointing at his elbow.

He looks at where she's pointing, where his skin has started to dry out and crack.

"Oh, it's nothing serious...I'm just peeling. I can compose a moisturizing cream with a little karité butter, aloe, and coconut or banana."

Nyala looks at him uncertainly. "There is only one person who knows so many things."

"Who would that be?"

"The Hogon," the twins whisper to each other.

CHAPTER FIFTY

Hogon

It is first thing in the morning and the men and women of the village are already active and about their business. The Pulldogs reach the main piazza of Sangha, where the *To-guna*, the Council House, stands, anxious to know the outcome of last night's meeting.

"There's no one here. They must have been at it until late…" Hakim says, not knowing whether this is a good or a bad thing.

The building is square, about four meters high, built with layers of pressed millet stalks. The roof is piled on top of a rough frame resting on three rows of parallel drystone columns. Instead of lintels there are anthropomorphic trunks carved to look like heads and busts of the ancestors.

"Why is it so low? You can't even stand up under there," Rafabel says.

"People come to listen and talk in the *To-guna*, there is no need to stand, or raise your voice. If someone gets irritated or agitated, they end up banging their heads on the ceiling," says Hakim, smiling. Then he gets serious again. "The word commands respect for the Dogon."

"These sculptures aren't wood though," Shimbo says, looking at the lintels.

"No, they are cellulose and lignin 3D-printed copies. The originals are hidden in caves in the escarpment. There have been thefts in the past…someone let themselves be bought off cheaply."

Then a crowd of people are surrounding the group.

"Happy, Nico? The essence from yesterday…" says Rafabel caustically, "…it must have spread with unpredictable consequences, or else the twins are experts in getting the word around."

Nicolas is overwhelmed by bizarre requests and submerged by business opportunities: there are people who want to set up a business making soft drinks derived from cola nuts, with stimulant effects, there are people who want to film a web series on Dogon cosmogony, or a virtual museum of artifacts found over the course of the centuries.

Alphonse pushes his way shouting through the crowd and drags the Pulldogs to safety with the help of two well-built young men.

"Cheers to anyone who is thirsty!" he says cheerfully.

"How did the meeting go yesterday, Dad?"

"We'll talk about that later. First I have to talk to Nicolas."

The Pulldogs are escorted to Alphonse's house. The two big lads stand guard by the doorway to stop any petitioners from getting in.

"Please, enter..." says Alphonse, doing the house honors. "Hakim, yesterday I didn't ask you to sleep here because I could see you wanted to be with your friends. Our house is very different now to how it was when you left."

The house isn't bare like the other Dogon homes; there are paintings in bright colors by African artists hanging on the walls, a wooden dresser and a four-poster bed draped with mosquito nets. There are also decorated vases which are being used to ferment millet beer.

Alphonse invites his guests to sit on the rug and offers them tea, the three-cup ceremony. After the ceremony, Rafabel half expects to see the family treasure chest brought out, or a suitcase full of masks, handed down from father to son over twenty generations. She is disappointed when Alphonse pulls a flowery cloth pouch out of the pocket of his *boubou*.

He opens it and pulls out, one by one, a number of pieces of jewelry. Some are so shiny they look new, an ebony bracelet, a ring with a black stone; others, on the other hand, look scratched and worn out, a decorated silver plate, a fringed handkerchief, a bone pendant.

He carefully spreads the objects out over the rug. They are simply but carefully carved and all the decorations seem to hold symbolic meanings that Alphonse does not explain but lets them sense from the emotion in his eyes.

"These belonged to my wife, Hakim's mother."

Despite the fact that a lot of time has passed since her death, Hakim recognizes the bracelet; and he too is hit by surprise as Alphonse hands each one to Nicolas.

"Please, sniff them…I've heard what people say about you. If there's anyone who can realize the value of a smell, that person is you."

Nicolas leans forward and sniffs deeply.

"Can you smell it?"

"Smell what?"

Alphonse becomes solemn. He stands, goes to the dresser and takes a vase, then he sits down and holds it tightly in his lap.

"Efia."

Hakim looks at his father startled, amazed.

"Tell me if you can smell her smell."

Nicolas is hesitant. Up to now, capturing the smell of a person has done nothing but bring him trouble.

"I could smell the objects and something vague…like the hint of an exhalation, just implied."

"Are you sure? Nothing else? Please, it's important. I have tried wrapping these things in cloth and oiled bandages, then I asked the village women for advice on how to extract the essence with karité butter and alcohol. I called the best *feticheurs* in Bandiagara to get them to squeeze the objects in the presses in the factories there, I even went to Bamako where they told me there was a Frenchman, an expert in cold *enfleurage*, but no one could do it for me. It would be a wonderful thing, a treasure that shouldn't be lost."

Nicolas looks at Hakim, then Alphonse; they are both hanging on his every word. He turns to Rafabel, seeking advice in her eyes. She gives him a small, kindly smile as if to say there would be nothing wrong in giving the essence of a wife back to a husband, or a mother to her son.

She winks at him and that is undoubtedly a sign that being kind to Alphonse could have a good influence on the final decision regarding the use of nanites.

Nicolas puts his face closer to the objects again. He inhales slowly. The wait is frustrating. From outside they can hear the voices of people who sound like they want to get into the house and ask for their own personal miracles.

"Really, I need you to leave the objects with me, so I can give it the time necessary. What you are asking is not easy."

"OK, I can do that."

Hakim relaxes and goes back to the reason for their visit to the *To-guna*.

"Now can we talk about yesterday's meeting?"

"Oh, we discussed it until very late.... But the matter was too complicated for the knowledge of the elders, so we made no decision."

"What do you mean no decision?"

"Nothing, we asked for the Hogon to intervene, only he can give us a sign. This evening, on the land beyond the southern altars, the divination will be held, and tomorrow morning we will have Yourougou's answer."

Alphonse gets the Pulldogs out quickly and goes back to his chores.

At dusk the light of the last of the sun's rays falls diagonally across the escarpment and on a man wearing a sky-blue *boubou* and a red beret on his head.

The man has reached Sangha by climbing a footpath twisting between sharp stones and is welcomed solemnly by the elders. He is one of the last Hogons in the region, custodian of the Dogon language, the umbilical cord of their society. It is this man who maintains the link between the land and the heavens.

He has traced a series of interconnected squares in the sand, a sequence of symbols representing potential future scenarios for the village: love and hate, health and illness, life and death.

"I can hardly believe it. People are dying of hunger and you are putting yourselves in the hands of this witch doctor?" Rafabel says to Hakim, trying hard not to get angry. He looks back at her impotently.

The Pulldogs have been allowed to watch the divination thanks to the intercession of Alphonse on their behalf. He has always fought for the Dogon rites to be shared rather than to remain mysteries, even if this means sharing them with tourists for a suitable recompense.

The Hogon places small sticks on various squares. The outline of the *Is* and the little piles of sand symbolize other worries: the harmony inside the village, next season's rains, the mortality of the herd, the arrival of a newborn, an economical problem.

"A Hogon only used to be a spiritual leader," says Hakim in a whisper. "A repository of the knowledge of the ancestors, the oldest man in the village who celebrated the most important events, presiding over weddings and funerals. Then things changed, the diviners became fewer, so a Hogon was also chosen according to his ability to read the divinatory signs, to be the intermediary between the visible and invisible realities. This is why we are outside Sangha at the moment, at the intersection between two universes, the human and non-human one."

"Do you believe in all this?"

The Hogon studies the drawings; he traces an imaginary route with his finger, moves some of the elements, dividing them as if to give emphasis to the shapes that have formed. He mutters to himself, rattling off his preliminary answers.

"I don't believe," Hakim answers. "But the invisible world and the way the Hogon manipulates those sticks remind me of the first time Nicolas showed me the heliotrons in Rome."

Rafabel says nothing; the memory is still vivid in her mind, too. Perhaps that was the moment she fell in love with Nicolas, with a man weighed down by too much flesh and fat, but who nevertheless believed in the immaterial strength of a perfume, and in the simple strength of an idea, like that of walking to revolutionize himself, and the world with him.

Nicolas is staring at the Hogon, watching his every move, as if the divination were nothing more than the oldest method known to man to shape the impossible.

When he is finished, the witch doctor scatters a handful of peanuts around the marks he has made; these will attract the foxes. Any prints they leave during their nocturnal wanderings will be interpreted by the soothsayer at dawn, together with any tracks left by the wind or the rain, insects or birds. The Hogon starts singing to invoke the sacred fox. The animal has to come to weave a path of prophecy. In a low voice, Hakim translates the words of the song for the Pulldogs:

Fox, please tell me,
Is there something?
Will there be hunger next year?
Fox, speak clearly,
Cast your tracks.
Will we feed on grains of energy?
Give me the claws to draw in the sand.
Whatever you see, tell me,
Give me your paw prints.

The Hogon's song fades like the last light of the day pausing in the sky.

"The night will send the fox to visit the signs." Having said this, the man disappears into the lanes of Sangha.

Rafabel takes hold of Hakim's arm. "You know we have to do something! We can't leave the future of your people in the paws of a fox."

"Of course, not least because no foxes have been seen around here for years, there weren't even any when I was small."

The Pulldogs are waiting two hundred meters from the place of divination.

"Everything is in the same place. Nothing has moved," Rafabel says, taking off her visor and checking what the others are doing.

Nicolas has downloaded a picture of a fox's paw prints from the Internet, and he is modeling positive molds of the paws starting from the negative. Farisa is busy programming the movement sequence to reproduce three different gaits of the animal: canter (overlapping paw prints), trot (diagonal pairs), and gallop. Shimbo is printing the drones necessary for falsifying the divination, or rather make it favorable to the answer they would like to hear at dawn. Hakim is overseeing the operation, afraid they might be discovered.

"When it walks, a fox only puts weight on four toes; the residual thumb has no function," he tells Nicolas, examining the details of the model. "It is like the print of a dog's paw but more elongated. I must have liberated at least six pairs when I worked at the zoo."

Then he moves on to check Farisa's work. Each of the four drones has been given control of a paw, they will have to use them to make the prints in the correct sequence to imitate the gait of the animal.

"Foxes move from place to place at a trot. The length of their strides goes from fifty to seventy centimeters, and the length of the paw print is roughly five or six centimeters."

Every now and then they go from trotting to a canter to make the simulation more believable. The final gallop will make it all the more difficult for the Hogon to dismiss.

Shimbo displays his devices with pride. "All I have to do is take the peanuts away from where the prints stay still for the longest."

"Good," says Rafabel. "We are going to change the future of the Dogon."

After mounting the prints on the frames, the walkers launch the drones and watch the divination through their visors, set to infrared. They imagine the silvery coat of the Yourougou appearing under the light of the moon and leaving its response and the future of the escarpment in the sand.

A few hours later, the first light of the rising sun reveals a shadowy trail left in the sand.

"It was here!" shouts the old man, barely managing to hold his excitement in as he examines the paw prints. "Yourougou has been to this sacred place and with his paws has foreseen the future of Sangha!"

The elders keep their distance and leave the Hogon to his task of interpreting the omen.

"I will tell you everything I see," he says slowly, as if possessed by the spirit of Yourougou.

The witch doctor notes the absence of the peanuts. "The fox has eaten the offering!" Then starts to intone phrases from memory that sound like ritual formulas repeated ad infinitum.

"The genies of the water follow you: if there is illness in your stomach you must cure yourself, if there is illness you must take the remedy."

More than watching the gestures and words of the Hogon, the Pulldogs are keeping an eye on the elders and their reactions.

"I have the news!" the shaman exults. "Someone comes with baggage, travelers! I see a union. There is no illness. We will have open heads, we will hold our heads high in front of everyone."

The elders start muttering. Alphonse makes a sign of approval.

"Yes, there is something in the way…something like fear, twisting our stomachs. But it is impossible to stop its progress. Long life. Laughter. Sacrifice: snakeskin and red cola nuts."

CHAPTER FIFTY-ONE

The Dama Mask Dance

The beating of the drums is a crescendo rising in the distance: from the edges of Sangha they can hear the compelling rhythm of a percussion orchestra.

"It's impossible," moans Hakim, who has been woken by the noise. He and the others have only just gone to sleep after their role in guiding the answer of the Hogon.

"It sounds like a party," Nicolas adds, stretching in his sleeping roll.

The volume of the *tam tam* increases, beating through the lanes accompanied by lots of feet stamping out the same rhythm. Shimbo goes up to the roof from where he can see the procession weaving its way through the village, wearing traditional costumes.

"Confirmed. There must be more than a hundred dancers zigzagging backwards and forwards. They are all wearing traditional costumes, even the masks."

"What?" Hakim yells, running up to the roof. Standing next to Shimbo, he can hardly contain his anger. "The masks are only supposed to be worn on three occasions! When an elder dies, when a Hogon dies, and during Sighi, the celebration for the star Sirius…" He jumps down off the roof and starts getting dressed hurriedly. "… and that only happens every sixty years. I've never seen it. In 2027 I was in Rome."

In a few seconds he's at the courtyard gates.

"Do you get why I left now? This is just another one of my father's ideas. I have to talk to him. Come and join me as soon as you can."

★   ★   ★

Not far from Sangha's central piazza there's a large area where a complex ceremony made up of rhythms, songs, and dances is beginning to take off. Hundreds of tourists who have come from the surrounding villages are sitting on logs set out in a semicircle. Many of them are nodding their heads and stamping their feet in time to the rhythm, some are filming and publishing in real time on social networks, and some are commenting on this unique and picturesque experience.

Even without Hakim's explanation the Pulldogs can see in the choreography, with the masks symbolizing men, women, animals, skills, and places like the sky and the afterlife, it is a representation of the world and its perennial march.

The dancers are wearing hoods with feathers and tall, brightly colored crests, others have wooden masks painted red, white, and black. There are representations of animals like antelopes, deer, birds of prey, and also five-meter poles representing multistory houses. The dancers advance in single file, imitating this variety of animals, then they make a broken line to mimic a serpent, and in the end they all jump at the same time, stretching out one leg and then the other, mimicking the tail of this reptile.

Farisa spots Alphonse and Hakim on the other side of the open space. They are standing facing each other, gesticulating, in the middle of a heated discussion. She makes a gesture telling the others to follow her as she skirts the area where the dancing is. As soon as they reach the pair, the cause of the argument becomes clear.

"…I've already told you, the tourists have paid to watch the masked dance of the Dama. We need the money to buy millet from the other villages."

"But it's a farce!" Hakim yells so fervently that his heliotrons become red. "Where are the bones of the dead person to be buried on the escarpment? And don't try to tell me Sirius B has come early!"

"My son," Alphonse says self-importantly, "why do you continue to refuse to listen? The tourists cannot understand the complexity of our rituals, nor would they like to live like us. But they are curious and ready to pay to come into contact with some aspects of our culture. So, seeing as some of them already know about Griaule and have read his books, the other elders and I have decided to make use of this situation as a common base for understanding. Griaule is the key for entering the world of the Dogon."

Hakim's disappointment is evident in his sneer, so Alphonse draws the Pulldogs into the discussion: "Let's see what your friends think. When Hakim was small nearly eight thousand tourists came here to Sangha every year, and when he left that number had risen to over ninety thousand, but now the earth is burning, food is scarce, and things are getting worse. Last year only three thousand people came through."

Hakim has no intention of giving in. Someone bumps into him from behind. He ignores it and counterattacks.

"You're talking like a marketing man. What about the eight baobabs planted in the center piazza to commemorate the ancestors? We used to hold the dances there, so they could watch."

"We moved things...there wasn't enough room there for all the dancers of the Troupe Nationale du Folklore of Mali," Alphonse answers, filling the words with pride. "These kids tour all over Europe."

Hakim can no longer control himself; he snorts as if the rhythm of the drums is overwhelming him. Young Askalu, tired of the arguing, joins in with the dancing. His precise and agile movements imitate animal ones; his eyes meet Rafabel's and he smiles happily.

"Seriously? Have you seen them?" Hakim says. "These kids are professional dancers and actors who are reciting parts, they aren't even Dogon people. Where do they even come from? Have you ever asked yourself if the wise men would approve of this?"

Alphonse lifts up the mask he's holding. He looks into the large black eyeholes and hesitates. "You're right about that, the dances don't carry the same significance, but the dancers know this and the gestures they make are the same ones; therefore we are in some way, even if you don't agree, conserving our identity. Without tourists the young people of Sangha would not dance, and in the space of a generation they would forget everything."

"Looks more like theater than a ritual to me," Hakim says and leaves, followed by Farisa, who is trying to calm him.

"Tourism is an irreversible social, economic, and cultural phenomenon!" Alphonse shouts after him. "The escarpment would be invisible and without value if there were no one to appreciate it! We don't do this out of vanity, but to survive!"

Nicolas doesn't know how to repair the fracture between father and son. Next to him Rafabel is encouraging him to seize this moment of confusion, because although on the one hand their presence could speed up an irreversible process as Alphonse says, on the other hand it could constitute a point of discontinuity that would help everyone.

"Alphonse, the nanites are what could help us all. The Dogon would no longer depend on tourists' money, or food from other villages…you would be able to return to celebrating your rites in full respect of the ancestors, like you always have done."

The line of dancers, holding fans and baskets, execute their steps to the beat of the bells and drums, stamping their feet, jumping on logs, climbing wooden ladders, linking arms and crossing legs with the more exuberant tourists.

"I hope so, because the prediction was clear. Even the Hogon had no doubts about the signs left by Yourougou. You see, here on the escarpment, we have always been alone. You might have noticed there is no sign of state services in Sangha, except for the odd poster ripped off the walls after the last electoral campaign. In the village there is just one school for all the children, and the hospital works when it's our turn to have a doctor from Bandiagara or Mopti. The water pump breaks down frequently, so we are lucky we can get water from the river, but this too is getting drier and drier with every year that passes. Last year we had a cholera epidemic."

"With your permission we would like to distribute nanites to the population as soon as possible. We could start to compose them right now."

"I appreciate your zeal...but we must wait for the dance to finish."

"How long will it last?"

"In the past, three days, but circumstances have changed, this dance will last until tomorrow morning at dawn. I hope you will want to participate in the nighttime dancing, it is very atmospheric with the fires."

Nicolas and Shimbo follow Alphonse, who has them sit down next to him on a bench carved into the rock. The elders, including the Hogon and Blaise, stand to greet them. Rafabel joins Askalu in a wild dance.

"Do you know what happens when two elephants argue?" Askalu asks, jumping around Rafabel.

"No, what happens?"

"The grass suffers."

Tired and dazed after an hour of music and dancing, Rafabel and Askalu sit down on the ground. Around them the fires light up the dancers and agitate the shadows in time to the percussion. Askalu yawns and lies down.

"Are you sleepy?"

"Yes, but I don't want to go to bed."

"Why not?"

"Because everybody else is still here."

Rafabel knows the real reason is different: there's no one at home waiting for him. The boy rests his head on his linked arms and curls up.

"C'mon, I'll take you."

He sits up and his mood changes. "Really?"

They are home in five minutes. Askalu lights a candle as he enters his room and then lies on his bed. Rafabel sits next to him.

"Is it true you and the other Pulldogs never eat?" he says, taking her hand.

"We eat sometimes, but it depends on what we do. If we run, we use up a lot of energy, we get hungry every four or five days. If we are rested, we might last as long as two weeks without eating."

"Thanks to the heliotrons?" he asks, following the intricate shapes moving on Rafabel's skin with his finger.

"Yes, you guessed right."

"Will you tell me how you got them?"

Askalu is intelligent. Every time Rafabel moves closer to him, he looks at her sideways as if he wants to show her he's not a little boy, that she can trust him and he would know how to look after himself in any situation. He is not very expansive, he is quick to laugh and joke, but he isn't pushy, nor does he ask anything petulantly or insist too much like other kids his age would.

"It's a long story, it started many years ago in Rome."

"Rome..." he whispers, dreamily. "Where Hakim lived?"

"Yes, why do you want to know?"

Rafabel is sure Askalu must have a plan in mind. When he speaks, he does so clearly and directly, no whining, no trying to ingratiate himself.

"Because no one ever tells me bedtime stories."

She snuggles in next to the boy and starts to tell him the story.

On the way back to Blaise's house, Rafabel is just reaching into her pocket for her printed copy of the keys when she hears two female screams coming from the courtyard. They are not shouts for help, but of fear, like you would expect on finding an insect on you or a bloody wound.

Rafabel runs to see what is happening and the sight that meets her eyes is petrifying: Nicolas is standing holding a phial that probably contains the newly copied heliotrons, like he told Alphonse. In front of him, nude and frightened, are the Diallò twins, their faces transformed by terror and disgust. Aissa, the braver of the two, bends to pick up a flap of skin from the ground; it is the size of a handkerchief and looks like it has been ripped off. It is a few millimeters thick, rounded as if it has come from a shoulder or a calf: a taste of Nicolas's molt.

"What's this?" asks Aissa in a whisper.

The strip of skin curls up on itself delicately. The heliotrons are still visible but vanishing slowly like burning card becoming ash.

"I told you, I'm shedding my skin."

Nicolas's legs are piebald. Patches of dark, worn-out skin alternate with smooth pink patches.

"This isn't peeling…this is transforming," says Nyala, shaken to the edge of horror.

"Well, you can see it like that if you want to," he answers.

They can only see it in a bad light, so bad they gather their scattered clothes and run out through the courtyard gate as if they have just encountered a monster.

Rafabel takes a few steps and looks at Nicolas, still incredulous, in his underpants.

"Let's do it like this, I'll not say anything. You start by telling me what happened."

After putting the phial back in its place in the bandolier, he leans against a tree.

"At the party I started feeling unwell, my temperature went up and I asked Alphonse permission to come home and lie down. I didn't want to be disrespectful, so I asked Shimbo to stay there with him as our representative, after so many molts I know how to manage the situation."

Nicolas scratches his shoulder; it is so pale it looks like he is affected by vitiligo.

"I drank a bottle of water and I slept for about half an hour, then I woke and started to compose, no, I mean I looked for some raw materials. The shedding started a couple of days ago and I wasn't worried about leaving bits of skin lying around."

He stands and walks disconsolately. Rather than dissipating his body heat, the movement helps him to think.

"Then the twins appeared, they said Alphonse had sent them to make sure I was all right. You saw what he meant by 'all right'. They undressed, Nyala said she was the 'medicine' and Aissa was a 'welcome present' from Alphonse. I said I wanted neither one nor the other, but they were heading towards me until one of them trod on that piece of dead skin."

Rafabel lowers her eyes to the flap of skin without saying a word; then she stares at Nicolas, wanting to believe him, even though a similar situation had already happened with Silvia, the truth of which only came out after Nika's paternity test.

This time, though, she can read it as clearly as if it were written on his face. He's no longer the man who needed affirmation from others to compensate an ego crushed by the weight of the deleterious relationship with his father. He's now a block of wood who allows himself to be sculpted and carved; every encounter over the last seven years of wandering has shaped him, sanded and smoothed his bark, thinned his branches and strengthened his leaves. All this wandering from one place to another has been the best way to be pollinated by other people's experiences and at the same time spread *ecolution* emanations as far as possible. In the same way Nicolas could not hide his molt from her, he would not know how to lie to her, nor have the motivation to do so.

Rafabel gathers together the strips of skin and puts them in the nanomat to decompose; she slides the nanite-filled phial from the bandolier and places it on the table. Then she moves closer to Nicolas and strokes his beard.

Her face serious and solemn, Rafabel lifts her skirt and sits on her man's lap.

"Now what, Nico? What's going to happen now? The twins will almost certainly report everything back to Alphonse."

"I refused his gifts, but I haven't killed anyone."

To travel is to scratch yourself, it is welcoming the unexpected, surrendering to the unforeseen, it is changing skin every time you need to.

"Judging by how they ran off, their terrified expressions, I wouldn't be so sure."

Then she lowers her knickers and guides him into her. Their breathing synchronizes, their humors mix.

To travel is to accept transformation, the discovery that no step is the same as another and every movement is different to the preceding one. It is knowing to know oneself: it is the only route leading to yourself.

CHAPTER FIFTY-TWO

Feticheur

When the roll of drums is heard again in the dreams of those who are sleeping and the remaining partiers lie scattered through the lanes, the courtyard gate is kicked open. Four men come in, armed with machetes, and Alphonse follows behind them.

"Wake up! Everybody get up!"

Hakim is the first to get up and react. The others rouse themselves, unconvinced.

"What are you up to? Is this how you treat guests?"

"Guests? I should have realized your return would have brought nothing but trouble."

"What are you making up now?"

Nicolas gestures to Hakim to let it go and faces Alphonse.

"It's the molt, isn't it?" As he speaks he opens his robe to reveal the advanced state of his skin change. A few heads peek in from outside the courtyard. The voices of curious people reach them as scared whispers. Rafabel knows those whispers are the symptom of something more serious: wonder – however extraordinary it may appear – transforms easily into horror.

Alphonse stretches his neck backwards, horrified. "Nyala Diallò came to me last night. She was shocked. She said you are not a human being...and now I can see why. She wasn't lying, she wasn't hallucinating. Those things you manipulate, those nanites, have ended up manipulating you...if you are turning into a snake, it is your problem, but you are disrespecting the ancestor because you want to turn into him, insulting his memory. This abomination, as far as I understand it, is not only happening to you, but also those who follow you and believe in this collective folly, including you, Hakim."

Alphonse orders his men to take the Pulldogs into custody.

"You cannot stay in the village. I'm going to call a meeting of the council immediately to decide what to do with you. In the meantime, you will have to stay locked up in the way house, outside Sangha, to avoid the possibility of contamination." His dry lips twist into a sneer of disgust.

Hakim tries to resist this arrest, attempting to wriggle free.

"How can you do this to us? We came here to help you!"

"We will find another way of surviving. We have always managed... progress doesn't only happen through nanites."

With two men holding him by the arms, Nicolas attempts a verbal defense.

"Wait, Alphonse, you talk about development using the metaphor of natural growth and apply it to society. It is true that organisms are born, grow, and develop, but you forget they go on to die."

Alphonse motions his men to let Nicolas talk, and he continues.

"You are forgetting centuries of zero growth when people continued to live and prosper anyway. The idea that the history of humanity can be represented by a growth curve is false! The desert peoples know this because they refused the plow and any other innovation increasing the use of agriculture. They remained nomads, despite the governments who wanted to make them become sedentary. And thanks to these choices they have faced climate change caused elsewhere, by other, industrialized economies. However, when people moved en masse to the cities, the farmers had to increase their agricultural production to feed them, and that was the end of them. The dream of economic development destroyed the Sahel. You and the other minorities can only save yourselves by going back to what you were...and the best way to do this is by using nanites."

Alphonse is not convinced. His gaze is indignant.

"If you put a scorpion in your mouth, you have to be careful where you put your tongue." Then he turns to his son. "It would have been better if I had not called you, you have simply brought us more problems."

They are allowed to gather their backpacks and then the Pulldogs are escorted out of Blaise's house and out of the village, accompanied by the whispering of the women and the wide-eyed stares of the kids who follow them.

"What does 'you have insulted the ancestor' mean?" Rafabel asks Hakim.

They have just settled down in the way house about a kilometer outside Sangha. At the entrance, two armed men are standing guard, their duty to keep the Pulldogs here.

"A Dogon myth. But it's ridiculous that he has taken us away from the village and shut us up in here."

The way house is a small mud-and-straw hut that was once used as a shelter when moving from one village to another on the escarpment. Now it is nearly only used by the more adventurous tourists who spend the night here before carrying on to other destinations.

"What is the myth? I want to know what I am being accused of," says Nicolas.

"According to a legend, one of the ancestors looked after the children while the adults were at work. One day he turned into a snake and scared the children, but when the men came back he had turned back into his normal shape and everybody thought the children's stories were just childish tales. But then it happened again, and one day the eldest son, coming home from the fields early, surprised the old man in the midst of his transformation. Shamed at having been discovered, he turned back into a snake and ran away. Chased by the eldest son, the old man hid in the cave of Kommo Dama and fell asleep. His pursuer didn't dare enter and waited at the cave's entrance. Hearing no more sounds, he was about to leave when he heard a roaring noise and a wave came from the depths of the cavern to break at his feet. Looking at the ground, he saw a stone left there by the wave, symbol of the alliance between men and ancestors, before the old man vanished into the other world."

Nicolas lays the bandolier out in front of him and starts to sift through the earth in search of useful molecules. "I'm sorry Hakim, but I don't understand what my crime is supposed to be."

"Insulting Dogon cosmogony. This is what my father has invented in order not to accept the nanites. The Dogon universe is based on two principles: the vibration of matter, and the perpetual movement of the cosmos. The seed of life is symbolized by fonio, a cereal with tiny grains, driven by an internal force capable of breaking through the thin layer they are wrapped in. You have been compared to a *feticheur* because you know how to manipulate those grains, represented by the nanites, and therefore split and reform the deepest bonds of matter. In the eyes of my father, the eternal spiral of the universe is like a Nanocad program."

"What is he afraid of? Real consequences or spiritual ones?"

"Neither, I think he has an agreement with some business or other and is afraid if they take the nanites the tourist development project will go sideways."

There's a lot of noise and hustle outside, then the guards throw open the door to let Alphonse in.

"You come with me," he says, pointing at Nicolas.

Nicolas grabs his bandolier, but Alphonse stops him. "Leave it here, you won't be needing it."

A little worried, Nicolas turns to Hakim.

"Where are you taking him?" Alphonse's son asks.

"It is no longer any of your business."

As soon as Alphonse and Nicolas have left, Rafabel starts scrabbling around inside her backpack. She pulls out a tiny drone and switches her phone on.

"Now we will find out."

Taking the drone from Rafabel's hands, Hakim slips it out of the window at the back of the hut. She switches on the connection and they see Alphonse and Nicolas's backs appear on the phone's display.

"There, let's just hope we don't lose the signal."

Thirty meters further on, the two join a group of three people and set off along a downward-sloping path twisting and turning alongside a precipice.

Alphonse slips between sharp stones and behind him Nicolas stumbles as he follows: he has been blindfolded and moves carefully, led by the other men who are helping him as if he is a precious cargo. Halfway down the cliff face, by a vertical fissure in the rock, the group stops in a wider space.

"I'll go first," says Alphonse, looking up at the black holes of the caves gawping down at them from the rock face. Wooden bowls with funerary offerings are scattered throughout the rocks. Nicolas is pushed along amidst wooden grave markers sticking out of the ground. He is made to sit down on a tree stump.

Rafabel watches the screen and imagines nauseating smells exhaling from burst abscesses in the ground. The air must have an obnoxious stink; it is so bad Nicolas puts a hand over his nose.

"You must experience these smells. In this way you will understand what I am about to offer you. There won't be another chance," says Alphonse as he removes the blindfold.

Sangha's cemetery stretches before them like a battlefield scattered with holes and ditches used as common graves and filled with bones and skulls sticking out here and there.

After a few moments of this torture, Nicolas seems to have realized the reason he has been brought here. Alphonse starts walking between the graves. Every now and then, he leans over to read the name of the deceased, carved by relatives in the wood.

"Missirì Cissé, a dear childhood friend. He was only thirty-five when he died of HIV."

He picks up a handful of earth and lets it fall through his fingers.

"Sonni Alì, we worked together as tourist guides for twenty years. Died of tuberculosis at forty-two."

Alphonse picks up another fistful of earth and turns to face Nicolas.

"We have a proverb that goes: 'Even if you get up at dawn, your fate got up before you.' I have to make sure the Dogon can survive on their own when the NGOs and the tourists and any other foreign investors have gone. Unfortunately, it always goes like this. Africa is like this cemetery, it is the grave of all projects. The international agencies and nonprofit organizations come here full of enthusiasm and good intentions, they initiate pharaonic projects, launch humanitarian missions, and then, when the funding has run out or they have no more volunteers, they go back to where they came from and we are left with nothing."

He stops in front of Nicolas and blows the dust into his face.

"Haven't the Chinese investors given you enough guarantees?" Nicolas asks, almost as if trying to provoke him.

"They are different, they have a long-term vision. Their settlements are not colonial outposts, they are coming to stay. However, at my age I have learned not to trust other people's money."

"What do you want from me, Alphonse?"

"Do you remember the proposal I made to you? To extract the perfume of my wife from her trinkets?"

"Yes, I haven't found a solution yet. These have been some intense days."

"If you can give me a repeatable secure method to create her odor, I could apply it to these people too, or rather, what remains of them. Even though the dead will still be taken away from the village and buried in cemeteries because of the decomposition of their bodies, they won't have to be pushed away because of their stink, they will be able to regain the dignity of having the same smell as when they were alive. Think of the positive consequences for the conservation of our culture. Think of the emotive value for those who will be our future ancestors. No one has ever thought about this before now."

"You want to start a business allowing you not to have to rely on anyone else."

"I want the Dogon not to be taken advantage of by anyone!" yells Alphonse.

"Me too, but here I am, held prisoner by the Dogon and accused of violating your traditions."

"I can help you with the council, if you help extract the essence of the dead."

"It is not the kind of exchange I can accept." Nicolas grabs the bandage hanging from his neck and puts it back over his eyes.

Alphonse clenches his jaw and looks darkly at the guards.

Rafabel switches off the camera in the drone and clasps her head in her hands.

"Why did he do that?" asks Hakim, shocked.

Nicolas's choice is bizarre, as much as Alphonse's proposal.

"I don't think it's masochism seeing how assured he was when he answered. Perhaps he wants to resolve the situation like he did in Rome, giving up his status. Perhaps Alphonse reminds him of his father. They have the same way of doing business."

Hakim stands and goes over to the window, holding his cupped hands out for the drone to glide into. "Then we are finished."

"I hate the Diallò twins." Askalu hides his face in Rafabel's lap.

"It's not their fault. They were scared by something they don't understand."

Sitting side by side on a log in Sangha's main piazza, Rafabel, Farisa, Hakim, and Shimbo are waiting for the council's pronouncement. Unlike Nicolas, sitting alone and motionless on a log carved to look like a chair, they have not been handcuffed.

The ceremonial drums beat out a solemn rhythm, slow and unstopping.

The Hogon comes out of the Council House with careful steps, wrapped in his *boubou*, dark and frayed by time. He helps himself with a twisted stick that reaches to his shoulder.

"The altar gives to man, and man in exchange gives everybody a share," he says, opening the ceremony.

When the Hogon reaches the center of the piazza he stops, looks at Nicolas, and announces the decision.

"By ripping off his skin, man violates an ancient prohibition. The skin is our protection given to us by the gods: losing it means losing their benevolence. This is why the man who violates the prohibition cannot stay amongst men, he has to die."

Rafabel cannot believe what is happening. All their proposals for saving the Dogon are about to vanish because of a banal misunderstanding. How is it possible that in addition to refusing their help, Alphonse has gone so far? She grabs Hakim's hand as if he can intercede with his father. Then she looks at Alphonse who, sitting amongst the elders, is wearing his official mask preventing any visual contact.

"But you, Nicolas Tomei," the Hogon carries on, continuing his circular walk around the condemned man, "will not die. You will instead atone for your guilt with a sacrifice, like the ancestor, to recreate a shattered alliance. The snake is superior to man because it possesses the nature of the *brousse* and is not forced to work. Animals are more perfect than man because they cannot speak."

Rafabel leans in towards Hakim and whispers in his ear, "A sacrifice?"

He shakes his head as if to say he has no idea. "The Dogon haven't sacrificed anything for centuries. It must be another of my father's inventions."

Animals are more perfect than men because they cannot speak. How should she interpret that edict? Rafabel thinks it might mean that speaking, a tool of progress and foundation, of social organization, is held by the Dogon to be a calamity. Was the power of speech released by the need to reestablish order over the primordial chaos? That if the world had developed without obstacles there would be no need for speech, nor technology, inasmuch as the two things are connected? Or perhaps it means that animals are more perfect than man because they are immune to human disasters?

"A sacrifice," continues the Hogon, "serves us and others, because its energy penetrates men, moves through them and transforms them. The sacrifice prevents the spread of contamination, focuses bad in one point and neutralizes it. This is how universal order is preserved."

The Hogon begins to stamp his feet on the ground; every now and then, he hits the ground with his stick in time to the drums.

It is the end. Rafabel can see Nicolas impassively accepting the course of events, sometimes he half-closes his eyes, sometimes he watches the elders as if there were nothing behind those masks. He is not afraid of this summary judgment, he is just waiting for the finale, whether it is tragic or epic.

Rafabel tries to attract his attention, moving her head and waving her arms to rouse him and push him to react and defend himself from this absurd sentence. However, those lowered, absent eyes, closed mouth, unmoving nostrils, make her understand Nicolas does not want to fight against the fate waiting for him. It is as if he has come far enough, as if he has walked enough and doesn't want to look for yet another shelter for the night. The continuation of their journey no longer depends upon him.

The Hogon stops before the elders, inviting any who want to to add something. Then – suddenly – Nicolas is asking for permission to speak, moving his head.

The drums stop. The Hogon comes over to him and squints at him as if Nicolas is perpetrating a huge violation of the ceremony. Then, the Hogon motions for him to talk.

"Periodic changing of the skin," says Nicolas, "serves to regenerate an ability that diminishes over time. In this way it's possible to transform but remain the same; it happens to flowers that have just bloomed, identical but different even after fifty springs on the same plant. Think about it, with every new season trees bring their vitality back into circulation with their buds, despite the trunk and branches having the experience of many autumns. In the same way heliotrons renew their strength under an old dermis with every molt. In the same way bare trunks become skeletons at the end of December and in March rediscover the impetuosity of new buds, the heliotrons consume themselves to transform into new skin."

"Good. Have you finished?"

Nicolas turns to the others present. If the authorities play deaf, perhaps the community will listen. If the authorities have decreed his end, perhaps the community can be reborn from his end.

"We came to Sangha because Alphonse Konerè told us about your problems. I can assure you it is infinitely more economical to feed people with nanites than through the growing, harvesting, stocking, preparation, consumption, and recycling of foodstuffs. These intermediate operations can all be removed, together. Rejecting nanites is suicide, economic and cultural suicide. Like traveling on horseback when everyone else is flying in airplanes. Every living being has two fundamental needs, nutrition and reproduction. The first of these pushes people to compete for resources and a fight between species, the second induces them to reciprocal cooperation and support. By removing the fight for nutrition using nanites, we will be left with mutual support."

The Hogon waves his staff, commanding the drummers to take up their rhythm again.

"Enough. You have said what you wanted. Your sentence has not changed, the Pulldogs must leave Sangha by sundown, but you, Nicolas Tomei, tomorrow morning you will be taken to the main altar where you will receive, alive, twelve blows of the hammer which will shatter the joints in your arms and your feet, then you will be buried up to your knees in the ground to reestablish the violated order. Your heliotrons will provide energy for the village."

When he has finished speaking, the Hogon starts to dance around Nicolas.

Altar receive his water and his flesh,
So his good strength does not vanish with his blood
So his bad strength is pushed away!

Rafabel feels someone's hand holding on tight to hers.

"I won't leave you. I'm coming too," says Askalu, tears in his eyes.

CHAPTER FIFTY-THREE

1kW Man

Rafabel wakes from a troubled sleep, wet with sweat and with the feeling of being watched. Nicolas is not there next to her. In the other tents, composed ten kilometers outside Sangha on the escarpment by Bandiagara, Shimbo, Farisa, and Hakim are resting. Unable to resist, she gets up to dry her back and drink a sip of water, but the uncomfortable feeling of being observed stays with her. She hears a noise, looks up, and recognizes Askalu's bright eyes through the air hole. He waves from above, sitting in a branch. "The sacrifice has been set for midday."

"How do you know?"

"Blaise told me this morning when I took him his water. He was sad."

"Go and wake the others. In the meantime, I'll get the drones ready."

It is nearly eleven when the buses from the neighboring villages unload hundreds of people. The Sangha's piazza is so crowded, the only places left are for people willing to sit astride the walls of the surrounding lanes, or climb onto roofs.

The drone is transmitting images; they are mixed in with the hundreds of other streams transmitted simultaneously by the tourist agencies, and this allows the Pulldogs to follow the sacrificial ceremony from their camp.

"I don't like it," says Farisa, troubled. "We can't just sit here doing nothing."

Rafabel doesn't let her eyes flicker off the screen, especially when the elders start to appear. A round of applause and a surge of murmuring accompany their arrival. Behind them, their helpers drag two planks of wood to the center of the piazza where a trench, almost a meter deep, has been dug. The men position the boards and nail them together to create a Saint Andrew's cross. The atmosphere is getting tense, the heat is rising too.

At midday, wearing the proof of his guilt – his bandolier – around his waist, Nicolas arrives and is made to climb into the hole. They remove his handcuffs and tie him by his arms and legs to the wooden structure.

He lifts his chin to sniff the air. Unlike when he was sentenced, he now looks around, curious to see how he could have possibly ended up as the center of such a bizarre spectacle. However, when he realizes he can't find the smell of his friends, his eyes close and his head drops.

"Are they really going to sacrifice him?" Farisa asks Hakim, clutching at his shoulder.

"I don't know," says Hakim. "When I was small my grandfather told me that animals used to be sacrificed on the altars throughout the year, leaving their blood to reinforce the bonds men persisted in wanting to maintain with the heavens. All kinds of Sangha's animals died on the altars, but it hasn't happened for a long time...."

"It's so stupid," Rafabel says with disgust.

"I know, and it doesn't matter whether it is divination or purification, for the Dogon a sacrifice has always had the purpose of redistributing the vital force amongst the people participating in the ritual, which for Nicolas is the equivalent of providing the village with the energy from his heliotrons, creating a circuit composed of the Hogon, the victim, the ancestor, and the population. This force recalls the Nommo in order to activate the symbolic function of the altar which receives the energy. There the fresh new force unites with what has been accumulated over the centuries during the ritual sacrifices."

"An energy reservoir...."

"Yes, the altar stores the force man must conserve and from which he can draw when he needs to. When the altar is fed, the Nommo comes to drink, to assuage his thirst and feed life anew. In the Dogon language, sacrifice derives from the word which means 'bring back to life'."

The live images are flooding the network and social media. The tourists – most of them ignorant of local traditions – think the scene is a tribal ritual like the masked Dance of the Dama, and Rafabel asks herself if Alphonse is satisfied with this Dogon culture promotional campaign. On Twitter the hashtag #deadmancharging is climbing the world ratings from China to Brazil: everyone wants to get their hands on a clear flow of free energy, and those present return the favor by transmitting the spectacle to anyone who can't watch it firsthand.

Farisa cannot bear it anymore and cracks her knuckles. "I've seen and heard more than enough. We have to put a stop to this absurdity."

"What do you think you can do?" asks Rafabel

"I don't know, but we can't just leave him there."

"If we try to intervene, we'd cause even bigger trouble, and I don't think Nicolas would want us to."

Nobody says anything until Shimbo looks more closely at the screen. Another two characters have arrived on the scene: a blacksmith with a mallet, and another with a wheelbarrow. Taking out a kind of battery, the man puts it down against the cross and hooks up two tapes ending in membranes woven with electrodes to Nico.

"After the ritual will there be day and night surveillance on him?" he asks Hakim.

"When they have broken his limbs, he won't be able to free himself or go far."

"We have to try to get him away. Whatever it costs," says Shimbo.

Farisa moves away from the group and begins to gather dry sticks. "I've had an idea but I'm going to need your approval, Hakim."

Amidst the columns of acrid smoke surrounding Sangha, an outline wrapped from head to toe in a long robe is moving in the opposite direction to all the other people. The inhabitants of the village are loading everything they possibly can on handcarts and wheelbarrows, beasts of burden and their own shoulders, to get away from the flames. Amidst all the shouts, running, and pleas for help, Rafabel is moving quickly towards the main piazza. Beneath her veil she's wearing a black anti-smog mask, the same kind as when she used to pull a rickshaw.

The fires were set at dawn, as soon as the village started waking up, burning things and empty homes, but avoiding people. From various points around Sangha, the Pulldogs first set the fires alight and then, after raising the alarm, helped coordinate rescue and assistance.

Rafabel keeps her eyes down, she doesn't want to be stopped, and pretends to be intent on gathering together her belongings when suddenly someone grabs her.

"Please, help me! My sister doesn't want to leave, she says it's her fault!"

Rafabel holds a hand on her mask to keep her face hidden.

The young woman looks like Aissa. Her twin Nyala was terrified by Nicolas's molt and now she is shut inside their house. Rafabel's heart is thudding in her chest.

Hakim wasn't happy about accepting Farisa's proposal to set fires as a diversion so they could save their friend, but he felt in some way responsible for their failure. Accepting Farisa's proposal and chasing his father out of Sangha at the same time seemed to him like an excellent way of making up for an error of judgment, and payback for the injustice they were being subjected to.

Rafabel shakes her head and raises her arms without saying anything to Aissa. She waves her hands in a gesture of apology and, holding tight to her mask, carries on to the cross.

In the two days Nicolas has been trapped here under the blazing sun, the people of the village have left cola nuts, berries, three snakeskins and some figurines. He's being burned to death. The water that has been poured into the pit to water his feet has been reduced to mud.

Hanging around his neck by a slim chain is a sign saying:

DON'T TOUCH – PANEL BURN

The last vestiges of nanotechnology in Nicolas's body have attempted to maintain his vital functions, but without the protection of his skin

– burned and exposed for too long to the wind, heat, and smells – he has become a grotesque caricature of a human being. His muscles are wooden and gasping for oxygen because his body has tried to defend itself by moving fluids to where he's overheating the most in an attempt to dissipate the excess heat. In those areas, his skin is scaly and wrinkled. The heliotrons are only mildly vivid under his arms. His forehead is covered in sweat and he's dehydrating fast. Within only a few hours he will be overtaken by shock and all his internal organs will shut down. The device connected to Nicolas – a modified battery – shows 1kW.

He tries to clench his fists. His fingers are working, whereas his wrists, in addition to being broken, are tied up with belts threaded through small iron loops nailed to the planks of wood. The same goes for his ankles. Then, as if something is upsetting the smell of burning, the acrid stink of houses and all their contents in flames, Nicolas's nostrils flare. Even in the midst of all this smoke, his nose can detect the fulsome and spicy perfume of Rafabel. Pietro's curse finally seems to be well and truly over. He opens his eyes.

Rafabel kisses and sniffs him.

"You shouldn't have come. If they catch you...."

"Quiet, we've taken precautions."

"You? I thought it was spontaneous combustion...."

"There was no time to lose. Can you move?"

"I don't think so. I won't be able to snap the straps in time to avoid the flames."

"I'll deal with that." Rafabel takes a knife from her robes and starts cutting at the ties.

Nicolas moans. Rafabel doesn't think it's out of pain, but because he knows he has made a terrible mistake.

"I shouldn't have let the twins see the dead skin. Misunderstanding Alphonse's intentions was stupid. The biggest mistake was stopping. We should have left the nanites and gone on our way."

Rafabel notices the scaling of his skin, the same skin that many years ago peeled from Nicolas the first time and made him a free man, is now becoming necrotic with him trapped inside. However paradoxical it may seem, swallowing a sip of water would not now help him; on the contrary, as Farisa explained, it would cause an edema of the larynx and therefore death by suffocation; his throat would swell up until it was blocked, preventing him from breathing. With a strong blow, Rafabel cuts through the belt and grabs Nicolas's arm before it flops.

"I wanted to believe...but I was fooling myself right from the beginning. I built an amazing sandcastle on unstable foundations. I imagined a solution, it's just a pity it is irrelevant for anyone who sees things differently...."

"Be quiet, Nico."

But Nicolas continues to mutter. She starts cutting the leather of the second tie.

"Hakim told me an African proverb, 'A stranger can look at a thousand things but will only see what he already knows.' Rafabel, I lived in a city until I was thirty-seven. The only nature I ever saw were plant pots, flowerbeds, or at the most, a park."

As soon as the belt gives, Rafabel supports Nicolas's arm and gently lowers it to his side. Then she swaps hands and speeds the rhythm with which she's cutting the third belt blocking his right leg.

"I was happy when I didn't know the route or our destination, crossing the unknown with no idea of what we were going to find. Now I know...."

The morning wind seems to want to help destroy Sangha. At this rate, if no firefighter drones come to intervene, the village will be razed to the ground. Rafabel wonders if Hakim will be more sad or happy about this outcome.

"I know the world has its laws and that frequently – more often than we think – these are not human laws." A thread of dark saliva is hanging from his lips.

"Do you have a metallic taste in your mouth, Nico?"

"I haven't had such a bad taste in my mouth for years."

"I'm nearly done."

Rafabel looks around. The shouting has moved into the distance, the fire is beginning to crackle around them. A red wall is advancing on them relentlessly. When she looks up, she can see a solitary drone watching them. She nods, knowing that on the other side of the camera Farisa is waiting for a signal. Rafabel lifts three fingers, the minutes she needs.

With his head resting on his shoulder Nicolas sees the battery's charge display: 1kW.

"It's ironic, on average a man burns sixty-five calories an hour while sleeping and seventy-five while awake. Walking burns two hundred, and you need one thousand six hundred and fifty to stay alive...every day. This body has allowed me to escape death more than once, to live thousands of experiences and to enjoy lots of relationships with friends and strangers, but above all it freed me from my sedentary existence and the uncertainty of western society."

One more cut and Rafabel moves on to the last belt.

Nicolas's dirty feet remind her how much he has become a fundamental element of her nervous system and how losing him would eradicate all the connections activated by his presence, by his odors, words, and actions.

She puts all her strength into it. Rafabel pushes and the boards shake. So does Nicolas.

"Done, you're free." She gives the drone a thumbs-up. "The village has been evacuated. Everyone has gone, and now it's our turn."

In the corner of the piazza, three hooded figures appear: Hakim, Shimbo, and Askalu. The first two are carrying a hospital stretcher taken from Sangha's clinic; the boy keeps looking around, checking nobody is going to stop them saving Nico.

Nicolas is picked up and placed gently on the stretcher like a relic.

Seeing his heliotrons are dull and almost lifeless, Rafabel makes a decision: Nicolas's message – that revolutionary one of 'nanites for everyone' he first declaimed years before – must continue to circulate, like poems handed down over the generations until they become myths, like secret formulas passed from hand to hand transform into cults, and experiences shared by many people, become custom and then culture. This is the only way Nicolas Tomei can survive, by encouraging his contribution to the human race to propagate through other people's nervous systems.

Rafabel pulls the bandolier from his waist and takes the damaged but functioning bracelets from his wrists. In a moment of lucidity, he manages to grab her hand.

"You read my mind. The fabtotum and the bracelet: take them to Nika, please."

"Should I tell her something?"

"The last phial, give that to her. It's a perfume I made for her. Then tell her, beautiful follies help us live better."

She's on the point of crying because this is the kind of legacy that makes people human. Nicolas twists on the stretcher; his movement unbalances the carriers.

"Does it hurt, my love?"

"Physical pain, the real pain is the absence of the people you love."

"Are you thinking of Nika again?"

"Yes, apologize for me, I haven't been a great father. But tell her what we have done. Tell her our home is the whole Earth, that the hills are its walls and the plains are its floor. Tell her every time we breathe, we inhale energy. Tell her we have wandered without a destination, we have muscles and the weather-beaten skin of people who use their legs and the sun to move from one place to the next. Tell her we have seen things we would never have been able to imagine from the Gianicolo Hill or a skyscraper in the EUR district."

"I will."

His nostrils open to breathe in every molecule of this moment.

"For seven years, you and I have woken up and gone to sleep together. For seven years, you and I have walked alongside each other, every day. I will miss your smell...."

Then Nicolas's eyes close.

CHAPTER FIFTY-FOUR

The Value of Future Things

Rafabel puts in her earbuds and starts listening to 'Roads' by Portishead, a song from her childhood, when the world was enclosed by the rocky cliffs of the Val Venosta.

It was her grandfather who taught her the best things. It was he who had pushed her to sell the house in Naturno and leave after the death of her parents. He had crossed Italy to pull her alive out of the rubble after the earthquake in L'Aquila; now the least she can do to honor his acts and Nicolas's life is to cross the world again to take Veronika her inheritance.

Rafabel has a young boy with her, so she's even more wary than usual. She has avoided villages and busy roads, she has stopped to sleep in safe places and eaten what the land had to offer them: herbs, flowers, tubers, and a little fruit.

Askalu is finding it hard to keep up: the nanites have been circulating in his young body for a short time and they don't yet allow him to maintain a faster pace. His body temperature, after the first few days of sweating and convulsions, went down to thirty-eight degrees, but he's still not used to living with that sensation of constant fever. He's strong though, and so is his will. He has not allowed himself to cry, only uttering the occasional moan when the nanites' replication hurt the most.

They took the bus from Mopti to Bamako and then called Moussa Korò who sent a pickup driven by Ismail and Abdhallai to take them to Nouakchott.

"Can you make it? The ship is down there, we'll have to get a motorboat to board."

Rafabel points to a moored boat. Askalu can't see it properly, but he knows it is one of those belonging to the Green Ark she has told him about, an extraordinary thing for someone like him who is about to sail for the first time in his life. Askalu nods and follows Rafabel towards a motorboat which a couple of men are pushing into the water.

They both turn, as if to say goodbye to the land that has been hosting them. Their farewell – perhaps a see-you-again – is also extended to the Pulldogs staying in Sangha, each one for different reasons.

Hakim wants to rebuild the village. Within a few weeks, a battery of self-replicating 3D printers will have recreated everything as it was. There is a risk that maybe Alphonse will come back to reclaim his place amongst the elders, but by then other elders, younger than the previous ones, will have taken their places. The new Sangha will grow on different foundations to the old one.

It was Farisa who convinced Hakim to raze the village to the ground. It was she who convinced him to rebuild it again. She will need rest in the coming months, her gestating belly already visible.

Shimbo, on the other hand, couldn't resist Aissa's charms. He had admired her from the first moment he saw her. Afterwards, he did everything he could to get to know her. After the death of Nyala during the fire, he stayed, for her, and to help Hakim rebuild Sangha.

Rafabel feels a tugging at her sleeve.

"I'm scared. I've never been on a ship like this. The ones on the Niger are little boats by comparison."

"You'll get used to it. It rocks a bit to begin with, but it passes."

"Like on the bus?"

"Something like that."

"When I'm on the bus, I always play the I-spy game. For example, from the window of the pickup with Ismail and Abdhallai I spied a pile of lamb heads, I spied how they print bicycle tires, and I spied two car fires. Then I stopped doing I-spy. If everyone driving a car or a motorbike took the bus, there would be less traffic, less pollution, and fewer crashes. When I'm big I want to create a community of people who drive buses, like Ismail and Abdallahi. Their trailer has everything, bathroom, kitchen, printers, and music."

"It will be difficult to play I-spy on the sea, but wait till you see Rome's rickshaws."

Askalu's face lights up. "Are we going to Rome?"

"Yes, one day, but first we have to go somewhere else."

"Where?"

"Here," she says as she opens iMaps on her phone and points at a blue circle.

"Lake Baikal, in...how do you say that? *Irzusk?*"

"Irkutsk, it's where our friends are going to be in a few months. Someone saw the video of Nicolas online and sent it to a friend of mine on the Global Walker app. She called me last night."

"What is your friend's name?"

"Silvia."

"Was she sad?"

"Very sad. She and Nicolas were good friends. They grew up together."

He doesn't ask any more, but goes back to the main point.

"How are we going to get to Lake Baikal?"

"Walking. There is a special road, you'll see it when we get to Russia, it's called a walkersway. In the meantime, we have to get to the Baltic Sea first. You're not scared of the sea, are you?"

"No, there is a Dogon proverb that says, 'If you want to arrive first, run alone. If you want to go a long way, walk together.'"

Their appointment with the Green Ark is like finally meeting a virtual friend in real life. Rafabel can see Askalu's excitement by the speed with which his eyes are darting from one part of the vessel to another. What she has told him about it has not been enough to dampen his curiosity: it's immense, luxurious with vegetation, damp, but cleaner than the lanes of Sangha. With its stylized tree flag fluttering high on the flagpole, the Green Ark is not how Askalu imagined it.

From the railings on the main deck, he waves Africa a last farewell.

"Will you miss it?"

"I don't know, but I wish I could have spent more time with Hakim."

"What about Alphonse?" Rafabel asks, wondering if one day this boy will thank her or hold it against her for having taken him away with his brother's, but not his father's, permission. He doesn't answer but rests his chin on the railings.

"He said I would be going to Bamako as soon as I was twelve. I would become someone there."

"Good, we are only a few months early," Rafabel says, trying to lighten the tension. "And you already are somebody."

"Even though we lived in the same house, we never spent much time together."

"Then the thing is to concentrate on the good moments."

Rafabel's voice breaks slightly, making Askalu turn to look at her.

"Why? Is that what you do?"

She touches her wrist and starts to rub it. Nicolas's death was worse than an amputation, the brutal, definitive, and irreversible suppression of the nerve, emotional, and affective input she received from him.

"Yes, memories remain and the past has to be enough for our future."

Rafabel can't help herself and sheds a tear for that part of herself that left her with Nicolas. There is no anesthetic for such pain.

They had buried him by a baobab, his epitaph carved into its bark:

NICOLAS TOMEI

ROME, 1989 – SANGHA, 2039

WALKER

The tree will gradually consume him, he will be pulled up through its roots and absorbed into the lymph feeding the budding flowers and filling out the fruit. Every year, Hakim will go and re-cut the letters, and every new season Farisa or Shimbo or someone from the village will collect the seeds and scatter them throughout Sangha. With time, the augmented seeds will give rise to other fruit and pollinate other trees and other minds, giving rise, perhaps to other solar plexuses.

Rafabel can't stop looking at her wrist. The bracelet feels like a terrible weight now. By touching a series of symbols, she composes the unlock sequence, takes the object off and hands it to Askalu.

"You do it."

"Are you sure?"

"I'm sure."

Askalu smiles from ear to ear. He takes the bracelet and with a powerful throw bowls it into the sea. Rafabel takes another bracelet out of her bag, it is Nicolas's. She puts it on.

"Can I use the fabtotum?"

"All right, but remember it isn't yours. I have to take it to a young girl as special as you."

"I'll be careful."

Rafabel takes Askalu by the hand and they head off towards their cabin.

PHASE SIX

ECOLUTION

'It is not the nature of tracks being tracks that should interest the walker, but the amount of grace, strength and perseverance with which we follow them.'
Margaret Atwood, *Oryx and Crake*

'Think of the fierce energy concentrated in an acorn! You bury it in the ground and it explodes into a giant oak! Bury a sheep and nothing happens but decay.'
George Bernard Shaw, *The Vegetarian Diet According to Shaw*

VERONIKA RUIZ-CORMANI

CHAPTER FIFTY-FIVE

Footprints and Seeds

Nika's breathing is light, a cadenced sound spreading through the Tunka forest. When touched, the laziest larches drop a fine rain of gilded needles to the ground. Beauty is everywhere, at every height, lying low down on the soft layer of white dotted with moss, mushrooms and lichens, scattered amidst thousands of tree trunks enjoying creating a natural maze, beyond the twigs and branches, above, in the blue sky with cotton wool clouds.

There are what look like lots of little faces pressed into the fresh snow, faces that are actually the prints of hares: their back paws make two elongated eyes, the front paws leave two almost parallel marks making the nose and an oval mouth. Those aren't the prints she is looking for.

Nika climbs over a fallen tree, slows down, and stops to study a sun-drenched slope. The warmth of the sun has melted the thin layer of snow and the wind has taken away what was left. Behind her, the sound of crunching snow warns her someone is coming. If she doesn't want to be caught up she'll have to move faster.

Where the snow has been compressed the prints are different. *What is pressed in the snow becomes relief and the concave becomes convex*, Nika says to herself, remembering lessons of Kenshij, her Path Master.

The trapezium shape comes from the structure of the hooves, the deep print and the lateral spurs. A wild boar's prints sparkling like silver coins minted by the sun.

Knowing she has found the right tracks, Nika starts walking again; in fact she speeds up to put some distance between herself and whoever is trying to follow her. Along the way, she stops occasionally to gather red and indigo bilberries hanging from the bushes.

The consistency of the ground is an illusion. Solid-looking mounds turn out to be spongy carpets or treacherous bogs where it would be easy to sink, whereas lichens that look damp and slippery crackle underfoot like dried wood shavings.

"Veronika, stop! Wait! Have you found anything?"

She turns, and in the middle of the larch trunks she recognizes a face with a sulky mouth. She doesn't bother to answer and carries on along her path. The person following her doesn't give up; he's fast and trots across the snow with long strides like a young reindeer.

"I asked you a question. Why won't you answer?"

"You know why, Vasja."

His tight, fluorescent yellow sports clothing highlights his muscles, particularly well developed for a twelve-year-old. He is, after all, considered the village favorite for the next ultra-marathon. Then he starts walking very fast, making wide circles ahead of Nika's trajectory.

"Oh c'mon…give me a clue, a solitary tree, a humpbacked hill, any kind of sign."

"If you spent less time composing sport suits and gadgets and spent more time in the forest, you'd know how to manage without having to copy someone else."

"My future is not in the forest. Every gram of my body is destined to run," he says and sprints ahead. "Not here though, on an athletics track."

"So go to your track, but you know just running isn't enough to win."

"You are so mean, no, you are jealous!"

Vasilij vanishes into the whiteness and Nika veers right. She was so taken up with the running, she hadn't noticed she was trampling the boar's tracks.

The Path Master is right. Walking alone is like studying, if you do it in a group you end up chatting and learning nothing.

After crossing a stream with a jump, Nika sinks to her knees and pulls a dried apple core out of the snow. She clears a three-meter area around the roots of a tree and finds another two of the fruit buried under the snow. She lifts her eyes and sees the bare branches of an apple tree. It's a specimen of 'muscovy apple', the only one of its kind able to survive forty degrees below zero. Its red-and-green-streaked apples – juicy and a little sharp – could feed her for months.

Follow the prints, find the fruit, survive. Task done. That bumpkin Vasja.... You don't win an ultra-marathon with just your legs.

After putting two freezing pieces of fruit in her pocket, Nika takes her backpack off, pulls out the nanomat and places it in the center of the area she has cleared of snow. Then she turns the phone on, selects a formula and attaches the reservoirs to compose a tent. Afterwards she scatters a handful of marbles over an area of a hundred meters.

"I don't like sleeping on my own."

"You're a big girl now, Veronika," says Silvia.

"But I'm still scared." She takes an apple from her pocket and bites into it.

"Of what?"

"The same old nightmare."

"The one in the city?"

As she chews, Nika nods at the phone and Silvia, on the other end, reaches out her hand as if stroking her through the screen.

"It always starts with a roar of engines, lots of lampposts, one after the other, concrete blocks, red and white tape everywhere, then giant tires screeching to a stop, and a car crumpling. The air is hot. Lots of white balloons exploding together as car horns blare. The driver of an SUV, dressed like a soldier, shouts: 'Move, move!'"

"It's not your fault if you had a bad dream."

"I know, but it was different this time. A man was climbing out of the wreckage and he wasn't hurt at all. He started walking in the middle of the traffic and the cars moved out of his way."

"What did he look like?"

"He had long hair, his beard was the same color as dirty snow. He waved at me to walk with him. That's when I wake up...."

"You've had a tiring day. Your tasks?"

"I found prints and seeds, like the Master wanted."

"Good girl. Did you come across anybody?"

"No, except for that copycat Vasja." Her phone beeps, recalling Nika's attention. "Mum, hang on a minute, one of the marbles has activated." She opens the app and watches the scene relayed to her by the proximity sensor.

Two figures are moving fast through the trees. One is tall and one is short. They are both wearing short-sleeved T-shirts. The pale light of the moon filtering through the bare branches of the pine trees gives perfect visibility.

The images show a woman with red hair in long dreads, and a young boy, about her own age, wearing sports shorts. Nika moves the cursor and the camera zooms in on the glittering marks moving along their arms, then they blend in with the scenery as if they were mimetic animals. The strangers shine with internal light, like a night mirage. In the time it takes for them to leave the perimeter, the marble deactivates.

"What was it? What did you see?"

"Two people. Maybe walkers heading to their nest. They had heliotrons. I'll send you a screenshot." Nika shares the file with her mother.

"Ah, those are the friends we were waiting for. You try to get some sleep. Daddy and I will be waiting for you tomorrow."

"Mummy, will you leave the cam on?"

CHAPTER FIFTY-SIX

Nomadelphia

It's almost day when Nika gets back to the nest, their mobile village. The camp stays the same over time, but changes in space; it leaves no trace of itself on the land, like the shadows of clouds and the perfume of wild strawberries that appear and disappear quickly.

Nomadelphia – the non-village where the non-residents migrate like schools of fish and flocks of birds – is not a place, it's the experience of those who live there in the period between two relocations.

There are a few who live in the trees, inside camouflaged globe homes hidden by the leaves, and a few who live underground, in cool, welcoming tunnels, but most of them are scattered through the clearing in exotic-shaped structures, inspired by famous paintings or futuristic designs.

Dozens of homes – high-tech shells, modular yurts, extendable stupas, by the side of conical teepees and spherical wigwams connected by walkways – will be moved in a few weeks to the banks of Lake Baikal. Then, after a pause for the seasonal ultra-marathon, Nomadelphia will head north to escape the heat of the Siberian summer.

Alan has taught Nika that home for them means the place where they sleep every evening. That home might last for a day, a week, or a month, but very rarely as long as a year. Being here or there is not important, what matters is the path they take, and above all the people they take it with. Actually, for him, the activities measuring out the passing of time, daily walking, the evening break, the nighttime rest (which in the summer become nocturnal walking, morning break, and daytime sleeping), don't identify places, but sequences of moments.

Since we live in a temporal flow, as we move we flow with time and it doesn't seem to hold on to us. When we stop, time doesn't stop flowing but we remain stationary.

"That is where the sense of boredom lies," Alan said to her one day. "In staying behind in relation to time which flows continuously."

"If I run...will I live more?" she answered.

"If you run, it feels like you are living more to begin with, it's true, but if you go too fast you can fool time and it doesn't like being made a fool of."

"What do you mean?"

"I mean in the end you pay for your speed with lack of control. You lose sight of where you are going, you end up losing yourself."

Nika trots along the slope and reaches the outskirts of the settlement. There is nobody about except for Vajra Sudha working in the vegetable garden. When he sees her he stops the drone's seeding, removes his hat and greets her.

"Good morning, Veronika. How did it go in the forest? Kasimir isn't back yet."

"It was fine. Your son is good, he'll have no problems with the exam."

The rhizomatic transhumance includes the cultivation of green areas for passing walkers, so anyone can find fruit trees and vegetables in greenhouses GPS-tagged on the Global Walker app. The only contribution asked in exchange are a few hours of work to look after the plants; whoever doesn't contribute is tagged, fined in enerbots, and in the end, if it happens again, they are expelled from the circuit which is managed through blockchain to prevent possible abuse of the system. Furthermore, for security from outsiders, the greenhouses are monitored by video surveillance marbles and defended by stun drones. To the north – in the Komi area – she has seen some underground ones, fed by artificial light.

"And the race? Will you be able to do it?"

"I hope so. Did you do it?"

"Me? No, I'm old. When Dikran invited me to join the Pulldogs the ultra-marathon didn't exist. The first edition was six years ago; you were too small to remember."

"You're right, but I know Tasia won."

"Oh, yes, I went to fish on the lake and I saw her arriving before everybody else. Nobody was expecting that. The first out of seven villages, before Pino even!" He starts laughing so hard his stomach wobbles.

"However, you mustn't worry, Mr. Vajra. The Master is training us well."

"...it's terrible!"

Nika hears Silvia shouting as she slides into the yurt. She pokes her head around the screen marking the entrance to the common room and sees Alan as he shuts down a video on his phone. Her mother has her hands to her mouth and, sitting next to them on the sofa made of sheets of cellulose, are the woman and the boy she caught a glimpse of last night in the forest.

"What's happening?"

"Nika, it's you. How did the exam go?" Alan hurries to ask her.

"Fine, it wasn't difficult, but what were you shouting about?" Nika asks Silvia.

"Nothing," she says elusively. "We didn't hear you come in," she says as she gets up to hug Nika. "Let me introduce you to Rafabel, a dear friend from when we were all in Rome. This is Askalu, a friend of hers. They have just arrived from Africa and brought us some news, some very bad news. Sit down, love."

The guests get up to greet her too. Nika notes her father is nervy and he is fidgeting with his phone.

"What were you watching? Is that what made mummy shout?" she adds, turning to her father. Alan turns the phone over and over in his hands, but he doesn't know how to answer.

"Well, are you going to tell me what's going on or not?"

The room is full of smells that Nika cannot identify. Underlying Silvia's habitual scent of linden, there's something acidic, like lemon, which seems to be coming from Rafabel, and other shades she cannot describe.

"That's why I shouted," her mother says, pointing at the phone. "I found out about it through Global Walker, but seeing it is completely different. It is a video of a person, well, he dies on live camera. It is very cruel, it is best if you don't see it."

Nika turns to Askalu as if asking him if he has seen it.

"I was there, I had to see it and I wish I hadn't had to," he says in English.

"It's over now," Rafabel says, trying to lighten the tension. "Actually, we are not here to bring you more sadness."

She opens the backpack beside her and pulls out three objects.

"These are gifts—" She stops, seeking a glance of permission from Alan and Silvia. "From us for your birthday. It's in a few days, right?"

"Yes, in three days, April ninth," says Nika, a little embarrassed.

"Here you are, we can't stay for the party because we have to leave the day after tomorrow to go to Lake Baikal to see your grandma, Miriam. So we would be happy if you could open your gifts now."

Silvia looks at Alan for approval. He dips his chin and then drops his eyes, as if he has accepted the decision unwillingly.

Nika unwraps the first gift and finds an object she knows well: a fabtotum like the ones she and her companions are learning to use with Dikran.

"Oh, thank you! My grades for 3D composition will improve with this," says Nika enthusiastically. She isn't great at assembling molecules; she doesn't have the patience to verify atomic stability and compatibility before composing something. Her approach is casual, often messy, at best 'heuristic' as Dikran once described her methods at the nth abortive attempt to create a pair of knee guards. She feels at ease in open spaces where she has freedom of movement, and when she has to deal with anything that slows her down, for example abstract thought or logical reasoning, she gets into difficulties. But when she's on the right track, more by intuition than conscious choice, her compositions are as good, if not better, than those of her classmates. She doesn't think she can come close to the beauty and complexity of the objects created by Rurik, who at twelve is already composing 4D nanites, calculating temporal development as well as the spatial one. He is a passionate fiddler and, what's more, Ivan Shumalin is his grandfather, and his father is Alexander, the inventor of the 'sunflower', off-grid devices installed along the walkersways, designed to stockpile solar energy and convert it into enerbots.

Alan stands and Nika's eyes follow him. There is something they are not telling her.

The second gift is a bracelet.

"This is a special object," Rafabel says to her. "It has seen oceans and beaches, it has crossed mountains and plains, it has crossed waves and tides to come here to you."

Rafabel puts it on Nika's wrist, fastens it with a click and switches it on.

"It will protect you and give you strength when you feel like you can't cope. You have the ultra-marathon in a month and this bracelet will be of great use to you. Look, if you select this, it gives you your position and over here you can set a track." Rafabel shows Nika various functions and Silvia goes to stand by Alan. "From here, with a special plug-in you can find what materials are available for the fabtotum to compose with over the mesh net, and here you can communicate with other walkers on Global Walker."

Nika's eyes dart from side to side frenetically, bewildered by how many places have already been tagged on the app. "Have you been to all these places?"

Rafabel nods. She would like to tell her so much and fulfill Nicolas's last wishes, but it isn't the right moment. Perhaps she will be able to tell her more in the future. For the moment, she has accepted, as Alan suggested, not to reveal the real source of the gifts.

"This is amazing," says Nika, enchanted. "It has a history too, right?"

Rafabel turns to Nika's parents. Cautiously, they approve.

"Yes, it's all there, perhaps some parts have been damaged, and others may have been corrupted by time, but I haven't deleted anything," she says easily. "It will be like traveling where I have been over the last seven years."

Alan rubs his eyes. His sweat has an intensity it doesn't usually have. Nika doesn't understand what is happening and moves on to the last gift. She ends up holding a phial.

"Should I open it? Is it a perfume?"

"Of course, this is the perfume of the places where I have been. I hope you like it."

Nika unscrews the lid and sniffs at the essence. Alan is staring out of the window and takes a loud breath.

"It's strange…" Nika says, after taking a deep sniff. "There are so many shades, changing from moment to moment. It's a smartfume, isn't it? Who composed it?"

"I made it. With this," says Rafabel, showing her the bandolier by her side. "Every phial contains an essence I have collected somewhere in the world. One day you too could become a perfume maker. Would you like that?"

"I don't know. I like playing music, and being outside."

Alan has a shiver of pride that turns into a little smile. Silvia manages to hold him back.

"What instrument do you play?" Askalu asks.

"The guitar. Do you play too?"

"Yes, I like percussion instruments. In my village I used to play during celebrations."

"So we can play together, but I don't have a drum kit. Rurik could compose one in a couple of hours."

"Oh, I don't need one, all I need are saucepan lids and glasses."

"Even I know how to make those, we can print them in five minutes. Mummy, may we?"

Silvia nods and the young people leave the room.

When Alan relaxes and Silvia hugs Rafabel, Nika sticks her head around the curtain from the other room,

"Ah, the person who died in that video, was he a friend of yours too?"

CHAPTER FIFTY-SEVEN

Dendrophoria

As usual, Vasilij sets the pace followed by Sonni Alì, the first is tall and well built, the second is small and fast. They are exploring the territory at a hard pace in their camouflage bodysuits. Right behind them, side by side, Jundra and Nika are following, wearing two-piece suits, vests, and shorts, their hair in colorful hair scarves. Behind them, Kasimir, the tall guy, whose stride measures about a meter, has to keep himself in check so that he doesn't overtake them all. Every now and then he gets distracted and ends up treading on Rurik's feet, who pushes him away huffily.

When Kenshij makes the signal, the class stops. They all form a circle and sit down in a grassy clearing.

"Today we are going to talk about history," says the Path Master, tracing a circle in the earth with a stick. "We start with a small circle. This is the history of our community. Who can tell us this story?"

Sonni Alì raises his hand. He thinks every lesson is a speed competition.

"We go to old people because they know the past."

"Good," says Kenshij, drawing another shape. "Now we draw another, larger, circle around the smaller one. This circle represents the history of the county. How do we react in this case?"

"We ask other walkers?" Jindra hazards a guess; she's only nine, her family can boast three solar plexuses: two brothers and an aunt.

"Yes, but that isn't enough because they might come from other areas."

"I would use Greenternet," says Vasilij cockily.

"Right, focused local research. If you use the Vegetable and Informational Net every five minutes to find out about some small thing, you will end up feeling like you have the answer for everything. But that information isn't really yours, you haven't learned it directly, you have only seen, heard, and read what someone else 'knows'; in this way you will end up *knowing* everything without actually *understanding* anything."

The master draws a third circle around the other two.

"Then we have the largest circle, general history."

Rurik raises his hand. "We have to observe nature. Man's history is subjective, natural history is objective."

The Path Master moves to a tree stump. The class follows and forms a circle around him.

"Well done, Rurik. The less we experience in first person the harder it is to trust other people's judgment. Trees, on the other hand, can help us, because they act like very reliable climate archives They have no ulterior motives and their concentric rings capture lots of information about the history of the environment, a little like plant barcodes."

With his stick, the Path Master strokes the stump and illustrates some concepts.

"This poor tree was cut down at least five years ago. However, a cut-down tree is like an open book. The irregular and non-centered circles are evidence of landslides, wind, and land erosion which make trees grow crooked and curved. Floods and fires leave marks on the bark and the exterior aspect of the plant. In the same way circles of ripples expand on the surface of the water, where each transmits to the next the energy of the stone causing them, every ring in the tree trunk influences those that come after it, even affecting the bark which then shows signs of ancient cataclysms, the memories of which are kept safe in the heart of the tree."

He runs a finger across the various rings, counting to twenty.

"Here, fifty rings ago the springs fed the streams and the damp of the rivers fed the trees. Earth and roots are soaked in the memory of the water, which with the addition of minerals became sap. The xylem vessels sucked the sap upwards and the wood assimilated nutriment and memories, and translated it into its language of knots, boughs, and branching."

He returns their attention back to the center of the trunk.

"These elements here, on the other hand," says the Path Master, referring to a sequence of rings, "tell us of branches broken by the force of a storm, trunks burned by wildfires, and tumors as big as melons blooming on the bark, twenty rings ago."

Nika raises her hand to ask a question. "And man's history? Dendrology doesn't include them."

"Men and trees have separate history. The history of humanity is known, a braid of politics, economics, and art, any virtual assistant can tell you during your walk, whereas what the trees have to tell us is much more mysterious. In any case, you can choose your branch of study after the race," says the Path Master and then he dismisses the class.

They all run madly and vanish into the trees towards Nomadelphia.

When she reaches the nest, Nika is careful to move silently.

Their guests have slept in the living room and are leaving tomorrow at dawn. Rafabel and Silvia are drinking tea and chatting. Alan and Askalu aren't there.

Her parents' strange behavior the previous day has made her curious, and, well, if they won't tell her anything she will find out for herself. She hides behind the curtain and listens to them talking.

"How are you organized? Is Alan still the leader?" Rafabel is asking Silvia.

"He doesn't have the time to be the leader, his music takes up a lot more of his time than it used to. He goes on tour a lot covering an enormous area, from Yakutsk in the northeast to Almaty in the southwest. Sometimes Nika and I go with him, especially in the summer. Actually, order between the Pulldogs is maintained by small groups of people through a temporary authority over shared software."

"You put yourselves in the 'hands' of an artificial intelligence?"

"Not totally, the AI is there to coordinate and inform everybody about the most important arguments, whereas the smaller decisions are delegated to the individual groups concerned. In this way we guarantee horizontal control, as we always have in the Pulldogs. It's like when we had those assemblies on the viaduct. Only now, because we are spread out over a larger territory, sometimes even over more than one time zone, we use the Greenternet mesh net. The AI is a kind of block administrator tasked to carry out our wishes. The concept of authority has never really stuck down our way."

"How are things with Nika?"

"While she was small everything was easy, she never doubted anything we told her. I was her tourist guide, her instruction book, her map and compass of reality. Now, though, she is entering pre-puberty and our relationship is going to change. I just hope she won't think of me as an enemy, I hope I won't make the same mistakes my mother made with me."

Nika jumps, Askalu has grabbed her tunic and is tugging on it to take her out of the nest. With the other hand, he puts a finger to his lips to shush her.

"Shh, this way. Your father has a surprise for you. He doesn't want to wait for the party tomorrow, he asked me to come and find you as soon as you came home."

"Why?"

"Come, you'll find out for yourself."

Askalu leads her to the edge of the forest, then he lets go of her hand and offers her a string. She holds on to it and notices the heliotrons on the boy's skin.

"How come you have them too?"

"Rafabel gave them to me."

"And how did she get them?"

"I don't know."

"Is that why you are going to my grandma? She has them too."

He shakes his head. "I'm sorry, I know nothing about that either. Rafabel hasn't told me yet. Listen, I have to get ready to leave tomorrow. But we'll see each other again when you come to Lake Baikal in a few weeks. I've heard there is going to be a special race."

"Yes, the ultra-marathon. We make a circuit of the lake, but for us it's more like an end-of-year school exam."

"Cool, can I run too?"

"I don't know. How old are you?"

"Nearly twelve. I'll ask my mother later. Now hurry, follow the blue thread. Your surprise is waiting for you on the other end of the string."

She looks in front of her and speeds up until she's running through the pines and cedars that grow tall and towering in the southern part of the Tunka National Park.

The string seems to go on forever. It skirts around trees, crosses ditches, snakes through bushes, and goes under heaps of leaves. Nika comes out at the head of a meadow sloping down for more than a hundred meters. Alan and Pietro are waiting for her at the bottom of the slope. One is holding the guitortoise and the other a bass. She races down the hill to where they are on the banks of the Irkut River.

An acoustic version of Keane's 'Somewhere Only We Know' accompanies her descent. Then she sits on the ground and listens to her father's rough voice.

As the song finishes, Alan moves closer to his daughter.

"Happy birthday, my love. Tomorrow you'll get presents from your friends and the other Pulldogs, but I wanted this to be my special surprise just for you."

Then he gives her a guitar, handmade and carved from the prized wood of a red spruce.

CHAPTER FIFTY-EIGHT

Born Runners

It took them eight days to cover the five hundred and forty kilometers to Shamanka Rock on Olkhon Island in the middle of Lake Baikal. Before the bridge was built, its two spans composed in six months by a group of transarchitects from Akademgorodok, you had to get a ferry to reach it.

The caravan did not go around Khuzhir, the most populated village on the island, to avoid adding more tourists to the procession. In fact, they allowed anyone who wanted to join their itinerant cavalcade as if it were a kind of unplanned, joyous, picturesque excursion. A little further south, in the winter, they run the Baikal Ice Marathon, between the villages of Tanhoi on the eastern shore and Listvyanka on the western shore of the lake. The race, one of the most extreme in the world, is held on the icy surface of the Baikal and attracts hundreds of athletes from all over the world.

In the sky a scattering of clouds, and on the ground a handful of dwarf pines with contorted roots growing on the stony slopes, give the landscape the flavor of an antique oriental painting.

If you continue to tread a path with your feet, some of that path enters into you, Nika thinks, going over the Path Master's lesson in her head.

In the distance, she watches the Nomadelphia caravan slowing down to approach the descent. On the horizon, a curtain of mountains like brilliant roughly shaped fangs glitter around the oldest, deepest, and largest lake in the world.

What you tread on leaves a mark on the earth and another on your feet. Any touch is an exchange with the outside world, Kenshij concluded during their last lesson before she left.

The first to arrive at Shamanka Rock, Nika and other kids are enjoying the spectacle of the panorama; the caravans are coming down from the hills, following the runners of the ultra-marathon. Waving her arms, she greets the people she knows from other transhumant communities like Neotopia, Wanderworld, and Noburgh. Then she gets up and starts on the last stretch of the road.

In front of her, Shamanka Rock looks like an alien meteorite that found itself unexpectedly in the clear waters of the lake. However, until a few years ago, the situation was different: the effluence from industry, the 'Baikalsk Pulp and Paper Mill', which closed in 2013, and the arrival of *spirogyra* alga had initiated a process of eutrophication that threatened the lake's whole ecosystem. The purifiers installed on the Selenga River to limit the pollution from the paper mills had not been able to compensate for the increase in urban effluence and nitrates from cultivated land.

The Baikal and all the creatures living off it had seemed doomed.

The *omul*, a fish similar to salmon, no longer let out their high-pitched scream when pulled out of the water. After swimming upstream to lay their eggs, they would return in November before the lake froze; in the spring the small fish just out of their eggs would be dragged to the lake, where their parents were waiting to eat them.

The Baikal, the old Buryats told her grandmother, Miriam, hosted many strange things like this. In its depths there is a crustacean called 'horse of the Baikal', which used to hold stones in its claws without any apparent motive, like ballast. Two-hundred-kilo sturgeons that took twenty years to reach maturity could carry up to nine kilos of caviar in their bodies. The minute *gammaridy*, shrimps with red eyes that lived at the bottom of the lake, amassed in their thousands per square meter, defying the dark shadows with very long antennae. In the abyss there were bizarre oily fish known as *goljanki*, so transparent you could read a book through them. The female would give birth to two thousand babies capable of swimming, and after doing so would float to the surface of the lake and float as if dead. Its body, though, once out of the ice of the depths where the pressure was high, exploded or melted, leaving puddles of oil and an ethereal spine, rich in vitamin A. The people of Buryat would use this oil to light their lamps.

The older inhabitants of the area told Miriam all these things when she arrived among them the first time, carrying a message as revolutionary as it was incomprehensible: 'nanites for everyone'. Now Nika's grandma lives where no other woman has ever even been allowed to go near over the course of the centuries.

Nika waves and greets her from a distance, as Miriam throws a ball in the air and performs dives and somersaults to catch it before it falls to the ground. At eighty-four, Miriam is still doing sport. As well as her morning Qi Gong exercises, she swims and runs for an hour every day. Her hair is gray and sweaty, whipping backwards and forwards in the wind. Her body is flexible and toned.

In the past, according to local traditions, women were forbidden from going to Cape Burkhan and entering the sacred cave of the shamans. For some, the ban was connected to the ancient belief that the presence of women, 'impure and sinners', could contaminate those hallowed places. For others, the ban was there to protect women, since it was believed that visiting the cave, full of mysterious and uncontrollable energy, could complicate pregnancy.

Her grandmother had broken all beliefs.

Her grandmother had given the lake back not only to the people of Buryatia, but also to anyone else who wanted to visit it. She had achieved this in a simple and innovative way. She had planted willows and poplars enriched with specific phyto-renewal nanites.

"What lovely hands you have, Nika, slender like reeds," says Miriam, holding her granddaughter tightly in her arms. "And the guitar? Is it new?" she adds, seeing the instrument hanging across her chest. Nika hasn't been separated from it since she got it.

"Yes, it is, a present from Daddy."

Nika's hair falls in curls over her cheeks like spilled tendrils, but the eyes looking out from behind that jungle are serious and concentrated, steady and unmoving.

"How are you? Are you worried about the race?"

"No. The race will be fine," she says without looking at her grandmother. Her curiosity has taken her eyes along the cave wall where an idyllic valley has been painted with natural pigments. It is a depiction of the birthplace of the legendary ancestor, and it includes some circular yurts surrounded by wild horses. The Buddhist monks, on the other hand, believe the cave is inhabited by a Mongolian divinity and often make pilgrimages to it. A few benches are covered with animal skins, and on the back wall an image of a shaman surrounded by light and flowers is stretching up towards the sun. The cave is a real covered sanctuary.

"So then, what is the matter, love?"

"They are," says Nika, pointing out towards Rafabel and Askalu's tent. "And you," she continues, taking her grandmother's wrinkled hands, on which the heliotrons are tracing faded patterns. "And then me. Why do only we have them? Since those two arrived I have thought about it a lot, but Mummy and Daddy don't want to talk about it."

"It's nothing strange, you get the heliotrons from me. Your parents don't have them because for them nanites are enough."

"Uhmm," Nika mumbles, unconvinced.

"That's how it is. Some people eat the fruits of the land, and some the sun's rays. In some way, every walker has embraced ecolution."

"If they didn't want the heliotrons, how come they let you give them to me?"

Miriam shows no surprise at the directness of this question.

"Because you aren't them. Alan and Silvia have spent most of their lives in a different reality, in different bodies."

"I don't know. You have convinced entire villages to use the heliotrons. Thanks to you, thousands of people have stopped eating like they used to. How come you haven't managed to convince them, too?"

Miriam smiles, and with a sign calls Lisa to her. Lisa is a Siberian cat with long white fur. "Every time I see you, I am always struck with how quickly you are growing up."

"I can't do anything about that," she answers, shrugging conceitedly.

"Right, it's useless to try to oppose the consequences of our mutation. I cannot complain about the slowing down of my aging, why should you complain about the acceleration of your adolescence? However, I didn't convince anyone on my own. I managed because there were lots of us, Ivan, Kirill, Kenshij, Shan, and other people who believed in Nicolas's ideas." As soon as she says this name, Miriam stops.

"Nicolas? The man who died in Africa?"

"Yes, him. What a tragedy.... But I was saying the people of the villages from Kransoyarsk to Yakutia, from Irkutsk to Magadan, were fed up of hearing the same old never-kept promises about the possibility of heating Siberia with nuclear energy, of irradiating the Arctic towns with artificial solar light, of thawing the permafrost and running the electric power stations with steam from the Kamchatka volcanoes, of building settlements with controlled microclimates under geodesic domes. They wanted food or something to replace it so they could be fed without too much work. Here agriculture is difficult, if not actually a risk. In a country like Russia that has always been crowded with spirits and local folklore, ecolution brought by the spread of heliotropism seemed to everybody to be a perfect synthesis between magic and pragmatism."

Lisa curls up in Miriam's lap; she pours a cup of green tea for her granddaughter and herself.

"Who knows, one day...maybe I will manage to convince your parents too. The Cossacks were the cowboys of the Russian wild east, the Old Believers were the Mormons of Holy Mother Russia. They lived spread out over a territory without frontiers, divided into various sects and groups even more isolated than the Pulldogs. But every community maintained a level of aesthetic rigor; they refused official baptism and any kind of church, they even looked askance at prayer."

They can hear the sound of the camp being set up outside. Askalu peers in through the entrance to say hullo, but he sees it's not a good time and Nika signals him to come by later. Her grandma is in a narrative trance.

"Some communities had started to baptize their children in the waters of the river and lakes and bury their dead in the forests. These

254 • FRANCESCO VERSO

sects wanted to escape the decadence of European Russia, and then of American Russia. In Siberia, though, where the authorities have always only held nominal authority, they were convinced the only way to salvation was through the soul. For them, history finished when the tsar abandoned religion and communism refused the spirit. Because of this, they chose to live outside all societies in a no-man's land where they would be able to be themselves. When we left Rome, for different reasons and without knowing it, we were feeling a bit like they were."

They both sip their tea.

"How many solar plexuses have you created?"

Miriam strokes Lisa.

"Ah, I don't remember! But before the plexuses, there were the trees. If we hadn't saved the Baikal we wouldn't have convinced anyone. There were so many solvents, hydrocarbons, and heavy metals in the water and along the shores of the lake that life was impossible. It is five years since the willows and poplars started growing again."

"I haven't seen any yet."

"That's because the last time you came, you were still small and they were only saplings struggling to survive. Nanites provided the solution. Kirill designed them to report pollution problems through bioluminescence, but then on the Green Ark, Nicolas turned them into a tool for renewal, making them capable of absorbing and transforming toxic molecules into gaseous chlorine, carbon dioxide, and water. Now the bioluminescence shows they are healthy."

"The Green Ark. I would like to travel on it one day."

"It's spectacular," says Askalu, who has reappeared in the cave. "I went on it on our way here."

"Don't think about the Green Ark. You two have a big race tomorrow and I have to make a speech to all the runners. I'll take you to see the luminous trees and then you're to go to bed."

Lying in the grass under the stars, hands behind their heads and eyes closed, Nika and Askalu are listening to the trees.

"It's those ones down there, the larches," he says, pointing to the left.

"Right, but those aren't larches, they are white poplars. You haven't been in the forest for long enough to recognize trees by their rustling."

Askalu opens his eyes; they widen and he touches Nika's shoulder to get her attention.

The wind is rustling through a line of birches, freeing a flutter of chrome-colored leaves. Something incredible is taking shape within the branches.

"What is it? Dust, or nanites?"

"I think it is pollen, nanites are too small to be seen."

Clouds of spores are floating in the air like smoke moved by the nocturnal breeze. Even these tiny granules have the bioluminescence of the plants that released them. The light fractures into a myriad of shards, glittering drops propelled by the pressure exerted by the trees.

"It looks like that painting, *The Starry Night* by Van Gogh," says Askalu.

"You're right! It's the same!"

"Do you know what nanites look like?"

"A little," Nika says. "Dikran, the nanotech teacher, told us how they have two engines for propulsion, a gyroscope, a growth battery, a calibration system, a cellular membrane sensor, and a manipulator. There are other parts too..." she pauses, trying to remember the names, "...a wireless aerial, a wavelength sensor, organized logic circuits, a T cell anti-receptor, and an emergency destruct system."

Askalu looks unwillingly impressed. "I bet you're the top of your class."

She blushes. "No, no, Rurik is the best."

Just above their heads, the clouds of atoms expelled from the leaves drift in gusts that smell of resin; it's like a shipwreck of bunches of inflorescence, a mass of transparent rarefied mucilage. Sniffing isn't enough for Nika; she raises her arms and lets her fingers touch the shoal of spores. Askalu copies her.

"Where I'm from, in Sangha, there is a proverb that says, 'Knowledge is a baobab trunk, one person alone cannot embrace it all.'"

Neither of them speaks. Touching the spores, taking deep breaths of the resinous smell, and measuring those fractal geometries, Nika feels like all of the science, beauty, and infinite wisdom possessed by the forest is passing through her, a kind of miraculous serendipity blooming unexpectedly when she least expected it.

"Do you think we will be able to communicate with them one day?" Nika asks dreamily.

"Who knows...plants speak through perfumes and chemical substances, butterflies use the ultraviolet, and dolphins use underwater sound waves."

"We use language."

"Yes, but perhaps communication between different species is possible."

"For me, nanites could become the interpreters of these languages."

Miriam appears behind them.

"Enough chattering. You are running tomorrow."

The dawn light reflects on the surface of the water, greeting the start of a special day. A procession is leaving from Shamanka Rock, crossing the fields and heading towards a large clearing. It is a little untidy in the way it proceeds. Boys and girls swarm excitedly; the adults discuss rituals that seem to be shared by various countries and villages to the extent they have replaced religious festivals. They have all been traveling for days and weeks to be able to participate in the ultra-marathon, and now the drones are proudly flying the flags of many communities: New Babylon, Podoland, Walkunism, 0 Town, and Freetopia.

In amongst the swarm of standard bearers, Miriam walks, wearing her red and turquoise tunic, her gray hair gathered into a bun decorated with colorful flowers. Her steps are accompanied by the stomping of her stick topped with a Baikal Seal head.

The winners of past editions are walking next to her, and behind her a crowd of young people dressed up in various styles follow; some have composed running suits with neck warmers, gloves, leggings, knee guards, and reflective jackets; some are wearing simple overalls with sweatshirts and trousers. Some of the girls are wearing makeup, some of the boys – not totally convinced by the whole thing – are distracted by their phones, others, the more anxious, keep checking the route on their phone screens.

As soon as Miriam stops, they form a semicircle around her to listen.

"Welcome to the island of Olkhon for the sixth edition of the ultra-marathon."

Her gaze embraces hundreds of people ready to celebrate this fundamental moment in their children's lives.

"Getting lost means the relationship between us and space is not simply one of domination, but also the possibility of being dominated. This is why changing places, facing different landscapes and a continuous hunt for our own points of reference, is regenerating on a physical and psychological level. This is why until you learn to be lost and find your way again, you cannot become an adult. This experience can be gained on the steppes, in the taiga, the tundra, and any other place suitable for transforming the normal state of our consciousness."

Askalu nudges Nika in the side. "Do you have your bracelet?" he whispers.

"Of course, but I won't need it. You just have to follow the coast and after two hundred kilometers you are back here at the starting point," she answers.

"Yes, I know, but we could chat a little in the evenings, when we are not running."

"Shut up a moment, let me listen...."

"Furthermore," Miriam is saying, "drinking water from the same fountain, sharing a night under the stars, washing together on the shores of a lake, gathering berries and seeds to eat, are all moments of communion that leave positive memories. To begin with, the children of walkers complain a lot about the strenuousness, some cry and refuse, more because of the mental effort and irritation than the physical trials. But then, once they have understood and discovered the pleasure of working their muscles in concord with their other senses, they thank us for having removed those mechanical and electrical devices that make the body a slave and the brain used to always choosing the easiest solution...."

The adults laugh, the kids don't; competitive tension is all that can be seen on their faces.

"Even our children will use machines and electrical devices, but they will do so after developing powerful antibodies; in our case these are called nanites and heliotrons."

Some of the parents produce water bottles with *tasarun*, a type of milk vodka they give to their kids as a kind of pre-race tonic. Others rub their children's quadriceps and calf muscles hard with camphor oil lotions.

"However, this is a competition, not a hike. You will be put to the test, you will have to manage on your own, you will have to solve practical problems and invent suitable solutions because running is an art and whoever runs is making an aesthetic choice more than a sporting one. Sport isn't the main objective, it is simply a collateral effect of movement. Pushing your limits along a route as hard and strenuous as this is the most gratifying experience humanity has ever been able to conceive. It is a splendid trial of maturity. The more I look at you, the more sure I am that the ripples created by the wave of your existence will give rise to many others...."

Miriam raises her hands to the sun.

"Let the race begin!"

CHAPTER FIFTY-NINE

Ultra-Marathon

Shoulders and knees low, steady strides, Nika's feet rise the necessary minimum to conserve the maximum possible kinetic energy. Sometimes she imagines she has wheels instead of feet, even though it would make the uneven terrain she has been running on for the last three hours even more tiring. Anyway, she hates wheels. Above all, pneumatic tires, the enormous, ridged ones that so often torment her in her dreams.

Behind her, Mount Zhima towers almost a thousand meters above everything else between eroded slopes where even the pine trees seem to be running in single file, downwind along the crests. At the moment, her objective is to get used to keeping her prolonged exertion under control. A ringing alerts her to a call. She could take advantage of it to slow down and take a sip of water.

"Hi...how...are...you?"

"Askalu, you shouldn't call when you're above the Van Aaken level," she says.

"The what level?"

"When your panting prevents you from talking easily. You're talking without finishing your sentences."

"Ah...sorry. Anyway...I've called you...now."

"Yes, what did you want?"

"A boy...has overtaken me...a black arrow...wanted to tell you.... Amazing."

"Thank you. Your body has more pressing things to do right now than give me race updates. Try to stay below this level, to save energy."

"All right...maybe you don't know...but in Africa...we just run."

When Nika sees the sun dip below the western curve of the rocks, the orange light of the evening turns blue. Every time they get in touch, she and Askalu avoid speaking using their phones. They communicate using their own sign language on the screens.

Closed fist = I'm fine and speeding up
Open palm = I'm fine but slowing down
Flapping hand = I need to rest
Clutched fingers = I need to eat and drink
Index finger and thumb making an L = I'm putting my tent up, to sleep

The vocabulary is simple and suitable for the circumstances.

Nika is camping in a dip in the hill she has stopped on. The race pauses between dusk and dawn and the runners have to stop for the night to avoid risks and dangers.

The moonlight held by the clouds turns the grass blanket into a chessboard with blotchy clearings and patches of vegetation. The horizon, which started light blue, has become dark purple, almost black.

The Buryat word 'Olkhon' has a double meaning when translated: some take it to mean 'little tree' (oy-khon), others prefer the derivation from 'olkhan' which means 'dry' because on the island there are neither rivers nor streams. To find out how humid the ground is, Nika digs a number of holes and keeps sticking two fingers in deep. It is as if she is a hygrometer more accurate than any divining rod. She has to move around a perimeter of hundreds of meters before she finds the best position to extract water; not spring water, but a solution dense with minerals. Then, she just has to concentrate, as if she is breathing in deeply, leaving her heliotrons to suck in the humidity trapped in the earth through the micropores in her skin.

Her supper consists of mushrooms, sunflower seeds, dried fish, raspberries, blueberries, blackcurrants, and redcurrants. The nanites in her digestive tract will extract the nutrients her organism needs, and if necessary, will transmit requests for integration by increasing her feelings of hunger. During the night, they will be busy, rebuilding her amino acid and vitamin reserves.

Before shutting her tent, Nika notices there are another four runners camped nearby. One of them stands out because he is in a fluorescent yellow sleeping bag, swaddled like a chrysalis. He isn't even using a tent.

The morning after, the western Sarma starts blowing, catapulting flurries of water against her tent. Nika sticks her head out: under the dark sky there is water everywhere, wind whipping everything and driving rain. A sudden storm is blowing branches along the path and vortices of stone chippings down from the rocks. Two bolts of lightning strike close to the tents; she can feel the heat of them sizzling through the air, there is a stink of sulfur, and the landscape is illuminated by a couple of violent explosions.

Nika is dazed for three seconds. With shiny eyes and aching head, she tries to reach the other tents. Water is dripping down her neck, smoothing her hair, plastering her T-shirt to her back. She is absorbing so much, she won't need to look for any for days.

The lake is so rough that it sounds like moans of fear are rising from its wind-whipped surface, and the temperature has dropped rapidly. It would be impossible to decompose anything in these conditions. Nika motions to the other kids to gather their tents, tie them up, and bind them to the trees, trusting to the slow natural decomposition of the materials.

They have to get moving and start running again as soon as possible.

Some of them already are. A figure is coming closer, driving through the rain and dashing between puddles. He is wearing a sport suit that is so close-fitting his body looks like it has been carved out of basalt. He is not wearing shoes, because two thick and shaped lengthenings extend over his feet, leaving prints in the sodden terrain like the treads of snow tires. When he passes in front of Nika, he gives a slight nod. A visor sticking out over his forehead protecting his eyes means he is not bothered by the rain.

He is perfect in his own part, she thinks to herself, going back to the soggy, muddy path. *I bet he trains like the turnik men: hard intelligent eyes,*

*dull suit like a black panther, and the cold, anonymous expression worthy of
a stone.*

For the following eight hours, Nika's strides seem to hit the
ground with twice the force her body weight warrants. Like
constant hammering on an apparently indestructible stone will
reduce it to powder over time, thousands of repeated blows during
the race will end up damaging her bones, weakening cartilage,
exhausting muscles and ruining her tendons. She doesn't feel bad
yet because her biology is not completely human, but despite this,
fatigue is creeping up on her. Knowing how much distance she still
has to cover, Nika slows down and chooses a place to set up camp
for the second night.

She finds a slope sheltered from the wind, sets up the nanomat, and
within a few minutes is crawling into her tent. After taking off her
shoes the smell of her tortured feet is so unpleasant it convinces her to
use the perfume Rafabel gave her. Just a drop to try to send her feet
away, down to where the ingredients for the essence were collected.

At her age she hasn't had much experience with perfumes. She
considers them a little like she considers films, where there is a story to
experiment with, or novels, where you do not immediately understand
the story because of all the shades hidden within. Perfumes, though,
have the advantage that all you have to do is sniff and they explain
something to you about themselves and what they contain.

Nika stretches her legs and lifts her feet into the air until they
touch the ceiling of the tent, and sprays a little of the essence onto
the soles of her feet. Reality warps and what enters her nostrils is
something incredible: expanses of green dotted with stone cairns;
wooden poles hung with strips of colorful fabric; herds of horses
grazing; the shape of mountains and immense cedars with drooping
branches in the background. She ends up dropping off in the mountains
of Nepal.

The next day, thirteen thousand strides further on, Nika notices a strange
rocky formation. It's the famous Stone of the Three Brothers, and a

Buryat legend her grandmother has told her explains its appearance. Once upon a time, three brothers lived on the island. Their father had supernatural powers. To grant them their wish to be free, he turned them into eagles, but only if they promised not to eat carrion. The three brothers promised to respect the wishes of their father, and happy with their new freedom as eagles decided to fly around the island. Despite the promise they had made, as soon as they were hungry, they found a dead animal and ate it. When their father found out he became furious and turned them into three rocks.

Nika runs past the stones and smiles at the thought; if somebody were to turn her into an eagle, she would have no problems maintaining a similar promise. A little further on she reaches Cape Khoboy, the northernmost point of the island, a stunning promontory, a natural balcony overlooking the Baikal with red rocks and seagulls.

"Hi, where are you now?"

"Hi, Nika, I've just passed a scary object, a kind of modern totem, made with off-cuts, car tires, mattresses, umbrellas, flasks, and lunch boxes. All stuff left behind by tourists."

"Are you sure it wasn't a rubbish dump?"

"Yes, the things were placed one on top of another, carefully positioned. As if someone wanted to leave a reminder."

"I'm coming up to it now." Nika lengthens her strides, rather than speeding them up.

"I'll wait for you then. We can run together for a little."

After a minute she and Askalu are running side by side, their knees synchronized, both of them toned, well-muscled, totally at their ease in the fullness of their exuberant youth.

If, while they run, the flowers scattered in front of them become an unfocused trail, it means they have come out of the aerobic stage. In those conditions there is silence except for their paired breathing. If, as they are running, they no longer care about the midges hitting their faces, it means they have reached the anaerobic phase. Once there, above one hundred twenty heartbeats per minute, the mind works in concord with the body.

In the evening, they camp opposite each other.

Before falling asleep, Nika turns on Rafabel's gift. The screen on her bracelet shows her unknown tracks, she reads exotic names of faraway places like Bandiagara, Bamako, Bomarzo, Büyükada....

And I'm only at the letter B, she says to herself as her eyelids start to droop. That night she dreams of an A to Z of fantastic places.

On the third day, things change. They start running southwards, with the mountains on their right. Askalu left a few minutes ago. The Black Panther might be in the lead. The other runners aren't keeping the pace, not even Vasilij, who every now and then appears and disappears in the hills like a reflective road sign, sad and alone.

Suddenly, after a curve, Nika becomes aware of something moving in the undergrowth. A muzzle pokes out from behind a tree trunk, then vanishes. Nika carries on running without paying it too much attention. The island is full of wild animals. Then, again, she hears a rustling in the bushes, an invisible presence trotting parallel with her. There are further hesitations, more kilometers, but in the end, the smell of human convinces it to come out into the open.

The presence is immobile, fifty meters ahead of her, testing the air. Stretching its neck, it sniffs. Its dull eyes look like crusted pearls. It is smelling her, investigating her chemical aura. For a puppy, the beast is enormous. A silvery stripe topping the dark brown fur curls like a ruff around its neck.

"Hello," she says, waving as she goes past it. "Where are we going to make our nest tonight?"

Every time Nika turns her head, it lowers its gaze, as if it were there by chance: perhaps it has already had enough to do with humans not to trust the first person it encounters. In that moment, Nika notices the dog's tail: it's strange, forked, like a snake's tongue.

Two hours later, the beast is crouched in the middle of the path, unmoving, as if waiting for her. Nika decides not to stop. She looks at it almost without breathing, she doesn't speed up so it doesn't think she's running away. After two hundred meters she turns, and out of the corner of her eye she sees it has moved, it is trotting, following at a gentle pace a short distance behind her.

Nika speeds up, and this time the puppy does the same. At a curve in the path, she turns for longer and gets a better look at it. It is a Caucasian sheepdog puppy. Its large head ends in a tapered muzzle, it is bowlegged, and then there is that bizarre tail, like twin Soviet sickles.

Nika has no time to lose, and its silence isn't helping; the puppy isn't whining, howling, or barking. No lament leaves its genetically modified body, and who knows why it has been altered in this way. Maybe if she starts running it won't catch up with her. Every now and then, when it doesn't keep up the pace and vanishes behind a hill, she thinks the dog has stopped following her, but then the forked tail reappears, wagging happily.

On the fourth day of the race, as soon as the mist dissipates, Nika recognizes the shape of the dog on the crest of a hill. The situation is different this time because it jumps and races down the slope as fast as it can.

Behind it, two thin, sinewy figures appear: wolves.

Nika starts running towards the puppy and the creatures stop in their tracks; then on the crest a third wolf appears and howls, a long and penetrating ululation.

Picking up a stick, Nika shakes it in the air as she cuts in front of the first two wolves. Usually when she does this they run away, but not these two. They sniff the air in her direction, puzzled; perhaps they can smell the perfume she is wearing, perhaps it excites them because of some strange combination of pheromones. What can she do if they decide to come for her? Her only option is to climb a tree, but what if she isn't fast enough, those beasts are fast. Then they back off. Perhaps it is a feint and they are waiting for their pack leader. It looks like they are going off to hide in the bushes, but then suddenly they are coming for her and Nika hears yelling.

The animals turn to look towards the hill. So does she.

Black Panther is running at them and, after a split second, the wolves are attacking him. The young man lets the wolves bite him, lets them try to plunge their teeth into his flesh, but his suit protects him from their sharp fangs, their wide-open mouths attempting to maul him.

"Run! I'll deal with them," he shouts to Nika and the puppy by her side.

The wolves bite, but get no satisfaction. Black Panther tires them out and they end up realizing this is not their lucky day.

Nika is running away as fast as she can, the puppy by her side.

"You're just a big squirrel, aren't you? Let's go. I'm going to call you Bielka."

Today is the last stretch, and judging by the presence of tourists in cars, she's getting close to Khuzhir. Here now are the relatives, friends, strangers, and curious bystanders crowded along the track leading to the finish line. Here are flags flying, people waving their arms and shouting encouragements under the generous sun shining down on them.

"C'mon! Keep going! You can do it, Nika!"

She wipes the hardened spittle from the corner of her mouth and the encrustations from around her nose. The air plays tricks at a certain speed.

Nika digs into reserves of adrenaline even she didn't know she had to drive her aching muscles on. Her knees go up and down but stay low; every now and then, she stamps down on the ground a little harder. She does it on purpose, to make herself react, to encourage herself.

She hardly lifts her eyes from the ground, her body is already running on autopilot, she is somewhere else, in the fantastic places she has seen in her imagination on her bracelet.

Black Panther didn't waste any time with the wolves, and is once more on the attack himself from behind. It will only take a moment's distraction for him to overtake her, it is inevitable. Part of her is overcome by panic, by the desire to put an end to this. It is useless, she cannot keep up the pace.

She can see Black Panther's legs milling angrily in the corner of her vision, he's staring at the ground too, but he's eating his own flesh, he's feeding on his own fibers in order to carry on pushing. He wipes some snot from his face with the back of his hand and he turns to look back at Nika; he's scared of losing too.

"Don't stop now! C'mon! We've almost made it!" yells Askalu in encouragement. He has appeared behind her like a hallucination; maybe it isn't even him, perhaps a part of her has called him up for the courage she needs.

Her suit is sodden. Under her T-shirt her skin is glittering and her heliotrons are vibrating convulsively. Her muscles have started to consume themselves, feeding on what is left, her own organic matter. She has never run like this before, without carbohydrates, emptied of all strength.

Where does energy come from? Where does energy come from?

Just one more kilometer, just one, every residue of her energy evaporates. She slows down. She grabs Askalu's arm and holds on to him to stop herself falling, just managing another ten steps before muttering, "Just a minute," and falling to her knees to vomit.

"Yes, get it all out," Askalu yells, "but hurry up!"

He turns to check out the situation. Vasilij's tired growl is ninety, perhaps a hundred meters away. Bielka barks, partly towards the approaching bright yellow suit, partly at Nika, inciting her to get a move on.

Askalu lifts Nika up by her shoulders while she's still vomiting, spouting liquid.

"C'mon!" And they start moving again, just about running. Vasilij is shining out sixty, seventy meters behind them.

Nika hiccups and pants, but she doesn't cry.

The spectators incite them, enthralled but shocked.

At this point Nika is only running on instinct, held up by willpower. They enter a funnel of people. More yells, more encouragement. She can't see anyone, not Miriam, not Silvia, not Alan, there's nothing else but movement.

Over the last hundred meters, Askalu doesn't stop encouraging her: it is he who crosses the finish line one step ahead of her, but she has done it too.

Black Panther won, but she is third.

CHAPTER SIXTY

Horizomantic

"I know it's a difficult choice," Miriam is saying as she exhales a perfume. "But try and put your rancor towards Rome to one side. I'm not asking Silvia and Alan the parents, I'm asking Silvia and Alan the people."

The two of them are looking at each other as if a boulder from the Shamanka rock cave is hanging over their heads.

"Nika is ready. She might be eight on paper, but her real age is different. She gave us proof of this two days ago," Miriam concludes, offering Alan the inhaler loaded with dandelion-flavored essence.

Rafabel's arrival has sped up a decision Miriam had been turning over in her mind for a while now. The death of Nicolas cannot be passed off as a normal bump in the road, nor can the symbolic value of the gesture for the Pulldogs, and all other walkers, be ignored.

Miriam wants to go back to Rome, she wants to do it as the person who has brought heliotropia to many people who never even met Nicolas; she wants to do this in honor of his memory, and because she wants to present a tangible, solid example of applied ecolution. She also has another aim: if Rafabel, who knew Nicolas better than anyone else, could speak to his parents, then she will be able to consider herself entirely satisfied. Not least in her desires is for her granddaughter to get to know her Roman origins, including her other grandmothers, Silvia's mother, Anna Ruiz, and Nicolas's mother, Olga Tomei, but she wants Alan's consent, despite his being against broadening their affective network.

Even though they seem to be a tangle, fused together so tightly as to appear inextricable, all these roots coexist in Nika and she should have the possibility of seeing them, as if they were various colors and substances, to find out what her past is made up of, and discover where parts of her developing character come from.

"I don't want to force you and I will respect all your decisions," says Rafabel, taking the inhaler from Alan. "But I would like to tell Nicolas's parents what I saw. I would like to tell them who he really was and for them to remember him as what he meant to us, and not just as what they would have wanted him to be."

"How long do we have to decide?" Silvia asks. She would not have been so generous and indulgent towards Petro Tomei. Ever since elementary school, she had never been able to stand his arrogance and pride, and if she were ever to see him again, she would not be able to give him her condolences for the loss of his son. For her, the closure of Il Romoletto was a positive thing, whereas her mother moving away from her home on Lungara was not. With her pension and some help from Silvia's brother in Australia, she now lives in an old people's home in the Aurelio district, though she would rather be back in Trastevere with a friend who is as alone as she is.

"Two weeks, enough time to agree who is going to come with us and organize our route on Global Walker. With the walkersway we'll need three weeks to go from Krasnoyarsk to Samara. From there we'll need at least ten days to reach Saratov and Volgograd from where we can board the Green Ark at Rostov-on-Don. It'll take ten days of sailing to reach Italy, skipping the marine blockades, and then the last leg to Rome. Two months in all, give or take a day. We were thinking of a goodbye ceremony for Nicolas. Not a traditional funeral, of course. The Pulldogs know how to celebrate special occasions," Miriam says, knowing she is touching her son's weak point.

Alan feels surrounded. He knows nothing about what has happened in Rome since they have been away, and just the idea of seeing the members of the old viaduct community again makes him feel a certain amount of what is almost nostalgia.

"What has happened to the Roman Pulldogs? Do you have any fresh news?"

It has been years since he last asked, years in which the present has been his only horizon in the morning, the only evening goal. No past, only the near future.

"The viaduct fell," says Rafabel sadly. She has kept in sporadic contact with Mario. "It was turned into a shopping center, like Nicolas predicted. The old General Markets now house the University of Rome Three."

The inhaler passes from Rafabel to Silvia. "And what happened to them?"

"House prices dived," says Rafabel. "Half the city is unrented. Three hundred thousand fewer people in five years. An unstoppable hemorrhage."

"Really? So many?"

"They are not all walkers. Some have chosen ecolution as a lifestyle or alternative form of downsizing, living in residential tent towns based on the concept of low environmental impact and direct compositive formula exchange. However, most young people simply continue to emigrate. They leave Rome because it no longer offers them any future. In some ways, the future is perceived as a geographical dimension that no longer passes under their feet because it has moved elsewhere. The Pulldogs stick with it and aren't doing too badly. They went back to the old farmhouse." Rafabel looks at Silvia: Serra Spino is the farm where their urban utopia originated, it is where she and Alan first met. "I can't wait to go back," she says, baiting the line.

However, it isn't easy to hook Alan, not least because it is obvious what they are trying to do.

"What I mean," he continues a little more softly, "is that I have never considered Nicolas as an enemy, because he wasn't. But I don't want to listen to all this shit about what a good man he was, clever, omnipotent, or how he imagined heliotropia and created the solar plexuses and the enerbots and the ergonet and this, that, and the other. I know you believed in him right from the beginning, Mum, what you have done with the lake and the local people is nothing short of miraculous, but I don't like folk sanctification, and I don't want them to turn Nicolas into a kind of Saint Nicolas of the Nanites. I've worked hard too, haven't I, to bring you here and organize our lives to be led decently?"

"I think," says Silvia, humoring his mood, "you have held true to what you said about the rhizomance."

"You all wanted to know the basics, where we were going, how we were going to survive, what we would become. I had to work something out! Anyway, everyone has always been free to create the future they wanted."

"Nobody is saying otherwise, Alan. You have invented a magnificent reality, in the best sense of the words," Miriam adds, taking the inhaler again. The cats start chasing the clouds of smoke, putting all the objects in the tent at risk.

"We followed you. Your ideas were more practical and accessible for a community like ours," Silvia agrees, almost as if trying to reassure him.

"Well, I could have tried to be clever, I could have sold it to you as a panacea for all the problems of modern life, sell you nature, freedom, independence, and this and that as different concepts to what they actually are.... I didn't, there are still problems, questions, just less absurd ones than when we were living in Rome. This is our reality now, we have to move and walk so as not to be eaten alive by entropy and the nanites themselves," says Alan fatalistically, almost resigned. "No, I am not a *brainy perfume and nanotech designer* like Nicolas. So it hurts me to hear you singing his praises every time someone says his name."

"He's part of our history too," says Silvia. "There's no point denying it, without him maybe the Pulldogs wouldn't be here, maybe you and me wouldn't even have stayed together."

"That's easy for you to say," says Alan bitterly. "You just have to carry on doing what you are already doing."

"Don't be silly."

"She'll find out," hisses Alan, landing a hard blow.

"So? Did anything change when the others found out?"

"It's not the same thing. The others are adults."

"She's big enough to understand that there are fathers and fathers."

"Fathers and fathers? What do you mean by that? What father would I be, what father was Nicolas?"

"Do we have to talk about this now?"

The discussion is heading straight towards a precipice.

"At least I took care to lie to her, to protect her," he says.

"Actually, you made Rafabel lie with that story about the gifts," Silvia says, holding her friend's hand in gratitude. "She backed us up, but now it is you and me who have to sort this out. It's time we faced up to it. I have lied to protect our daughter too, have you forgotten?"

He sighs. "That means we've both made mistakes. Sometimes lying is more difficult than telling the cold hard truth."

"Are you proposing to tell her everything?"

"I don't know. I'm saying I need your help," Alan says.

Silvia takes the inhaler, takes a deep breath and says, "Lies follow the path, you have to search in the grass for the truth."

Nika is running across the grass, following the sound of music. Bielka is almost invisible, only the track she leaves in the grass shows where she is.

She's feeling well again, the willow and poppy infusion made her sleep for a whole day, and in the meantime the wounds on her feet have almost healed. She can still taste the sweetness all over her from the honey Miriam applied to ease the aching. She lost two kilos, now partially reintegrated with liquids and minerals, but her body still hasn't stopped burning calories, consuming adipocytes even while she sleeps. Nika imagines the nanites as tiny, industrious ants living inside her, continuously making and destroying vitamins, proteins, and carbohydrates. There are many phenomena, from the most grandiose, like the movement of the planets, to the imperceptible, like molecular reactions that have nothing to do with our desires. The truth is, in addition to the millions of bacteria per gram of matter of her organism, a hundred trillion nanites call her body home. They adapt to what she does, they respond to her requests for strength and stamina; for them, she's nothing more nor less than a world to live in, to look after for their own survival.

When she gets to the top of the hill, she pushes through a crowd of waiting people. The first stars have come out in the clear sky over the island of Olkhon. Nika takes a peek at the picture she drew last night on her phone, and continues.

"Have you seen my father?" she asks Tasia and Pino, who are there for the concert too.

"Look backstage. Pietro has a temperature and Alan is auditioning the reserve bass player."

The Pulldogs' concerts are organized in two ways: like 'Burning Man', with Alan, who, as if he were an entrepreneur, puts together a band of musicians for a limited time and then they disband, or there is the other model which is like 'Cirque du Soleil', a kind of wandering circus performing a different but recognizable show each time. Usually, the tours follow three monthly migrations, during which Alan lives only for music and by music. While the fulcrum of that migratory orbit is represented by his presence on the stage, he doesn't care where he's going to play next, nor how many people will come to his concerts. His anarchic musical soul is enough for him.

Khuzhir is the first stop of the Horizomantic tour.

Nika finds her father sitting on a tree stump holding his inseparable guitortoise. He and another young man are practicing a cover of 'Yellow' by Coldplay. Alan gestures with his chin for her to sit next to him.

Alan lifts a finger and points to the object Nika is carrying on her back. She pulls the guitar around, holding it with a certain amount of shyness, and begins to pick out the notes of the song while her father carries on singing. When they play together, Nika feels a little bigger.

At the end of the song she moves closer to Alan and shows him her drawing.

"Yesterday evening, Mummy told me about the journey. I don't know if I want to go to a city...."

He meets his daughter's eyes for a moment, then looks back, staring at the picture she has drawn.

"Why did you make the cars into monsters? Are they so horrible for you?"

"That's how I see them, they are there every time I have a nightmare."

"In the city there are bad things, but good things too, especially in Rome."

"I know, grandma talks about it a lot, but I'm scared. People will laugh at me, they'll say I'm strange, that I live like a savage, they'll beat me up."

"What are you saying, Nika, nobody is going to beat you up."

Alan takes his fabtotum and connects it to his daughter's phone.

"I'll tell you what we'll do, it helped me when I was small. Maybe it will help you too."

He opens the Nanocad and starts to print her drawing.

"Will you go and get some bits of wood please?"

For seven years, since he started touring all over Asia, Alan has dreamed about changing the world with music, one song at a time. One night, under the effects of too much inhaled smartfume, he even confided to the Pulldogs that he saw himself like a singer on a decades-long road concert. A musical endeavor never before attempted in the history of music, not even by the Rolling Stones, Queen, or AC/DC. For how many more years would his enhanced body allow him to carry on playing? Ninety? One hundred? It didn't matter if the concerts were on the plateaus or in the forests for hundreds of people instead of in stadiums crowded with thousands of fans. He saw himself as the founder of Gerontorock, for music lovers over eighty or one hundred.

Then, during a concert in a park outside Bukhara in Uzbekistan, something extraordinary had happened. The makeshift stage had been set up in front of a video-projection installation representing the mausoleum of Ismail Samani which was destroyed in the earthquake of 2031. It convinced Alan to organize the remaining dates of his tour to be wherever a monument had vanished from, whether due to natural or human causes. It was similar to Nicolas's idea, he realized this, but there was a significant difference: he didn't want to convince anyone of anything, nor did he think he was anyone's savior. A series of concerts wouldn't change the world: "Bravo, wonderful, we should do it again here, and there," was the general level of the comments he heard repeated again and again. Like the video-projection on the destroyed mausoleum, a concert tour lasts for days or months at the most, whereas monuments challenge time, their strength lying in their ability to transcend the human temporal scale and leave a mark that lasts as long as possible. By uniting rock and destroyed monuments, Alan would make rock monumental.

After a minute, Nika comes back with a bundle of sticks which Alan arranges as a campfire.

"You light it."

She pulls out a flint and after making three or four sparks, she manages to produce a little flame. In the meantime, the car drawn by Nika has gone from the 2D of her phone's screen to the 3D of reality.

Basically, Alan's idea was simple: find a way to scan the destroyed monuments, the ones that would be forgotten and never seen again, and reprint them, life-sized, giving the past to the present and future generations.

This was how the Nine Domed Mosque of Balkh came back to the world, and the Babri Masjid mosque of Ayodhya in India, which the High Court of Allahabad had divided between Muslims and Hindus after its destruction in 1992.

If everything goes well, that is, if with their concerts they reach the computational capacity necessary to run the 3D printers, they would even be able to recreate the Buddhas of Bamiyan, blown up by the Taliban in 2001. Despite international efforts to rebuild Afghanistan after the war, neither the Japanese government nor other organizations like the Bubendorf Institute in Switzerland nor the ETH in Zurich have managed to do anything yet because of missing authorization and slow bureaucracy.

For Alan, the possibility of technical reproduction, nullifying the costs of the original creation, would defeat once and for all any attempts at destroying art, which would then always rise again, like a phoenix from its own ashes. From that point of view, the Horizomantic tour is not a simple concert, but above all a 'tool of mass construction'.

Sometimes he finds himself wanting to talk to Nicolas, even one of those discussions that ended in arguments and fists. He has told his daughter everything about it, he doesn't want Nika to imagine a situation different from the reality. Even though to face a truth like that of her paternity will take more courage than he thinks he has.

Alan watches the tongues of fire and winks at his daughter.

"Good, now take the car and destroy it. Smash the car and you will destroy your nightmares too."

She hesitates, unconvinced. It seems too easy as a way of getting rid of a fear she has always had. "What did you do it with?"

He would like to say so many names, of companies, of people.

"If the past makes us better, we have to preserve it because its presence makes us stronger; in its shadow we grow and become ourselves. If the past makes us worse, it has to be destroyed because its presence weakens us; we do not grow in its shadow, nor do we become ourselves." Alan takes Nika's hand. "I did it with a car, too." Then he guides Nika's hand holding the object over the fire.

Bielka yelps. Perhaps she can feel the delicate tension of the moment.

"I was small, and I was stuck inside a Mercedes. I couldn't open the doors, so I turned up the volume of the stereo just before fainting from the heat. They saved me just in time. I had nightmares about it for years."

Nika opens her hand and drops the car into the fire where the flames lick it until it catches.

"You have no reason to be scared of the city. We will be like seeds arriving from far away, only seeds that travel so far manage to germinate. Those that fall near the home trunk will not grow and flower well. Shan Jao, the dendrophora, says it better than me, 'The plants that become yellow far from the tree that seeded them will be the ones that live on.'"

Together they watch the fire turn the car into cellulose. Alan hugs his daughter. "How do you feel now?"

"Better, thank you."

"You were born on the sea, Nika. You don't have a passport nor citizenship. It's normal to be scared, but trust me, traveling helps to regulate our imagination about reality and to see things as they really are. That bracelet is a fantastic gift," he says, taking her wrist between his fingers, "but traveling yourself changes you inside, deep down. I won't leave you to go on your own, you, Mummy, and Grandma. I'm coming with you too."

"What about the tour? It has only just started."

"And it has already finished. I have unfinished business in Rome."

Nika slides out of Alan's embrace and puts her feet back on the ground.

Her fear hasn't vanished, but she's beginning to feel the sensation of there being infinite paths opening up around her.

"Is it true all roads lead to Rome?"

CHAPTER SIXTY-ONE

Walkersway

The Pulldogs are moving again. Above a natural stage, unchanging for hundreds of kilometers, heavy clouds are hanging, oblongs tinged with orange and blue. Some are clean and light like tissues fluttering in the wind. For Nika it's like an immense open-air cinema. She's enchanted by the ever-changing moving shapes: she sees herds of flying hippogriffs, pirate ships attacking galleons, plummeting spaceships, a castle under siege, abstract sculptures, floating tower blocks, and strange faces, happy, funny, sad, cheerful…a slow, infinite sequence migrating across the sky, like she and the others are migrating across the land.

A path shines where it winds through the trees. Alongside it, lines of birch trees, enriched with photophore nanites containing luciferase enzymes, emit an ochre-colored light indicating the direction from one stage of the way to the next. Those shimmering auras take her back to the lessons she's missing. Shan Jao, Kenshij's wife, is their actual Path Master, though it would be more correct to call her dendrophora inasmuch as they are on a walkersway, in particular on the first and oldest track created in Russia, almost five years ago.

"The plant realm alone represents over ninety-nine-point-five percent of the planet's biomass," Shan Jao is saying to the kids following her. Her oval face has the high cheekbones and olive skin typical of Kazaki Mongolians. "The animals – including human beings – are only trace, between zero-point-one and zero-point-five percent. Despite deforestation, plants are still the queens of the planet. This is the only reason why life on Earth is still possible."

Nika has already traveled along similar tracks when she was a baby, but this way is special. Her mother told her the first ones were created spontaneously along channels that had been cut into the forest originally to install high-tension cable pylons. The first walkers used them to move from one village to another, hardly ever coming across an asphalted road. Others came into being along abandoned railway lines nature had already repossessed; still others were created by regenerating roads that had fallen into disuse because of faster alternative routes like motorways or the low-pressure tubes known as hyperloops.

The walkersways are never crowded, and between Krasnojarsk and Novosibirsk they are used by many different ethnicities. The faces are Mongolian, Slavic, almond eyed, with Tartar cheekbones, Greek noses, and an infinite variety of somatic traits: Caucasian, Siberian, Baltic, Arab, Indian, and who knows how many others Nika wouldn't be able to recognize without a scan or mesh net search. So much variety can't even be found hundreds of kilometers to the south on the New High-Speed Silk Road connecting Xi'an to Istanbul via Urümqi, Almaty, Samarcanda, and Teheran. Perhaps down there the same ethnicities travel – for business reasons – masking their differences beneath gray suits, striped ties, and leather briefcases that are all the same.

"Our biology," continues the dendrophora, "not only derives from monkeys and dogs, but also from begonias and oak trees. A forest must be considered as a distant relative; she too reacts to the light, sends signals for reproduction and possesses many of our genes. For thousands of years plants have followed a different evolutionary path, but we have a shared genetic past. Heliophoria is the attempt to group together the family of species that use photosynthesis as a method of sustainable survival."

Proceeding along the walkersway is a truly enchanting experience. They meet people with packs on their backs, time wasters who are loafing around waiting for inspiration, freelance researchers exchanging formulas with strangers, groups of explorers without a destination; a flow of happy and mixed people; families traveling to far-off relatives, singles in search of bucolic adventures, old people enjoying nature. The Trans-Siberian Railway used to come past near here before the new generations opted for faster means of transport like cars, planes, and high-speed travel tubes. Now the railway, forgotten symbol of an industrious and manual Russia, is a dead-end track. Vice versa, heliophoria has pushed various communities to rediscover the abandoned paths that within a few years have transformed into shared habits: no walkersways are created in isolation, and since the route is planted with heliotrons and – like ancient postal stations – at regular intervals there are recharge *sunflowers*, some people call them 'solar brick roads'.

In Russia, like in other places, the governments tolerate them because they see them as a useful tool for thinning traffic coming into and leaving the big cities, even though in reality these tracks tend to avoid towns as much as possible, a phenomenon the authorities have neither considered nor foreseen. As with other tendencies in the past ascribable to the subcultures of the masses – such as raves, jogging, veganism, social networks, and animal therapy – the mayors of local governments have never been interested in learning more, only really interested in keeping public order. On the other hand, for them, a growing number of walkers helps improve health statistics and lowers the cost of health care, valid arguments for any electoral campaign.

"Who amongst you has heliotrons?" Shan asks the class.

Askalu nudges Nika in the side, she's distracted by all the faces around her. "Hey, she's talking to you...."

"She could actually be talking about you, too," Nika shoots back. He shrugs.

"I do, dendrophora," says Nika loudly.

Shan Jao turns and slows to walk by her side. "Hi, Nika, Kenshij told me we would meet. Have you gone through your first molt yet?"

"Not yet. Mummy told me it will happen soon, maybe this summer. Then after four years, and then it will stabilize depending on what I do. No one really knows exactly...."

"Because you are the first person to have them from birth. Later you and I should talk about the trees in the city. Your parents have told me where you are headed."

"Anyway, he has heliotrons too, and he's coming with us," says Nika, pointing at Askalu.

The dendrophora greets the young boy, who is intimidated.

"All right," she continues, "come by my billet this evening at the nest."

Askalu has the face of someone scared of being interrogated.

"That's enough for today," Shan tells the group. "Remember the koan of Path Master Kenshij: the walkers are on the road without having left the nest; they are in the nest without leaving the road."

"I don't know..." says Alan, scanning the ground with his smartphone looking for materials to compose his tent. "Is there any sense in still teaching kids how things work in the big cities? They will never be in one for more than a few days of their lives."

"You are making the same mistake the city teachers made with us: everything was theory and we didn't learn anything about things outside the urban environment," says Silvia. She's selecting a rustic design from the e-Den database that blends in with the colors of the taiga along the Ob River.

"Nika was crying yesterday, do you know why?"

Silvia is shaking her head as he gets a positive alert on his screen, a notification of compatible material, and prepares the phials for collection.

"Because the dendrophora told her that in cities the average lifespan is fifty to sixty years shorter than ours."

"Did she also say this is because they don't have nanites?"

"I don't know, I hope so," Alan says, filling the phials. "She used her phone to search for pictures of old people in cities, she saw how they live, how they spend their old age, how they end their days in isolation, without help from their families or the community. She became sad and then burst into tears."

"I still think children should know about these things," says Silvia. "They need to know about everything, including the negative things and those that are different from our principles. Have you forgotten all the magnificent things there are in cities?"

"Like what?"

Alan hooks the just-filled phial up to the fabtotum.

"Art galleries?"

"You can see them in 3D Vivo anywhere," he says. He motions for Silvia to start the composition.

"Historical monuments."

"They can be copied."

"Right, like the ones you are recreating?" As she says it, Silvia modifies the coverage of the tent in real time.

"Exactly," he says.

"It's not the same."

"It's close enough. C'mon, what else?"

She thinks about it, looks at the screen, thinks again, but says nothing.

"Why though? Why take this knowledge away from them?" is all she ends up saying.

"Because it makes them suffer," Alan shoots back. "For them it's inconceivable to live how we used to live, shut in by the four walls of a studio apartment, in a basement, or suspended on the tenth floor of a tower block. It was for us too, but we had no choice, we were born there and we adapted to that inhuman situation. Then, of course, telling them about rush-hour traffic, the bureaucracy surrounding everything, the instability of work, the shopping centers, obesity, anorexia favored by advertising, hospitals where you need to have private insurance to be treated, ADHD, and antidepressants.... It would be like a horror film for them."

"Are you sure it would be bad for everybody?"

"Well, I'd rather be wrong. The only exception is to go with them, like we are, so we can give them help if they need it."

Pairs of dragonflies fly over the grass that slopes down to the shores of the Ob. Some of the kids are still swimming, while the adults gather fruit and seeds for supper. The sound of the water is a pleasant gurgling mingling with the chirping of cicadas hidden in the trees casting shadows in the early summer. Alan and Silvia enter the just-composed nest.

"Another thing," he says, lying on the mattress. "When I saw she was crying, I thought maybe the moment has come to stop protecting her. We have to let her go."

"Go where?" The note of protectiveness slips out. Silvia takes a look outside. In the middle of about twenty tents, she spots Nika's, a rainbow-colored, segmented dome composed next to Askalu's.

"I mean in general, let her make her own experiences, alone."

The sky is lowering, it's like a dirty gray ceiling, the compact clouds a uniform sheet from east to west. The clouds accompanying the Pulldogs no longer look like cauliflower and aubergines like a few days ago. Sometimes walking just means putting one foot in front of the other.

The kids follow Shan Jao for the daily peripatetic lesson: she knows the taiga, she gives them advice on where to sleep – discarding the sometimes inaccurate Global Walker reviews – and knows where to find the seasons' produce. This is her walkersway, and she's the embodiment of this anthropic route. Her husband Kenshij goes even further, he maintains this 'spirit' must be protected in a formal way, not only conceptually, and he has been fighting a legal battle for months to obtain recognition as the legal representative of the walkersway. A little like the Kofan people of Singapore had acted years before regarding the Ecuadorean Amazon. They won a lawsuit to protect springs feeding the Aguarico River, nullifying fifty-two mining concessions and liberating thirty-two thousand hectares of rain forest.

"Take this tree, for example," says Shan Jao to the kids, stepping closer to a cherry tree. "During pollination it produces pretty white flowers that are perfect for attracting bees which cannot see the color red. The cherries turn red, but not for them, for something else. Does anyone know what?" the dendrophora asks, showing them a pair of the ripe fruit.

"For the birds!" the kids chorus back.

"Well done, the color red stands out against the color of the leaves and is visible from a long way off, even to birds flying above. A bird will eat the cherries, stones and all, go back to flying and drop the stone in its feces somewhere else: an efficient system of transport, good for the tree that has scattered its seed far from the mother tree and for the animal which has eaten. But attention!" They all turn to look at her. "The cherry will only turn red when the seed is ready; until that moment it's invisible to the birds thanks to being green, hiding in the leaves."

"Why?" Sonni Ali asks.

"The same reason all parents protect their children until they are ready. This is why an unripe fruit contains toxic or unpleasant-tasting substances. It's the plant's way of defending its young from predators before the seeds are mature."

In the distance, at the end of the path, they can see a group of luminous stripes coming rapidly closer. Every now and then, they fluctuate from right to left, gliding along a few centimeters from the ground.

"What are those?" Sonni Ali asks again.

As the figures come into focus and become recognizable, they can see they are a group of 'sunners' surfing thanks to the solar sails pushing them along. On the lead board, surprising everyone, there is a face they know, Ivan Shumalin, with his brother Andrej right behind him.

"Hello everyone!" he says, waving his arms. Once he gets off the board, he explains why they are there. "We have come from Kursk, for a Walk Pride. Some days ago, I spoke to Miriam and we found out about your pilgrimage to Rome. We want to join the expedition in honor of the memory of Nicolas Tomei. We all feel a debt to him, for how much he gave, and how much he left behind."

Miriam runs to hug him. After their exciting escape from the Green Ark, they have met up every now and then when he has been able to take time from composition projects of other vessels to come and see her for a few weeks in a summer camp here and there.

Their greetings and hugs are warm and sincere. Then Ivan becomes serious, and hugging Rafabel to him, tells her what he can't say in words: mourning, loss, compassion. In the end, seeing Askalu is curious about the board, he takes him gently by the shoulders.

"A solar surfboard like that is tempting, isn't it? You must be Hakim's brother. You have come so far to be all the way up here."

He nods, intimidated by the presence of this man with silvery hair.

"My mum helped me."

For a moment Ivan is perplexed, then he turns and understands. "Ah, Mummy Rafabel, of course. She can give you excellent advice, although I have heard you prefer driving to walking."

Askalu blushes. "It's not true!" He denies the accusation, waving his hands. "I want to drive a caravan, but as a public service to reduce pollution and traffic."

"Pollution and traffic?" says Ivan, smiling. Looking around bewildered, he opens his arms wide. "But here there is nothing of what you speak!"

"In the cities, that's where they need me," Askalu explains.

"Oh! A boy with clear ideas."

Then Ivan greets Alan. Their eyes meet meaningfully, and Ivan feels Askalu tugging on his sleeve.

"Do you have the formula for the board?"

Ivan unlocks his phone. "*Konechno,*[16] it'll be here in a moment."

Pino coughs hard. His eyes are red and full of tears.

"Well, then? Do you recognize the composition? Time is almost up," says Tasia in a singsong voice. The countdown is about to end.

Tommaso and Pilar snigger, lying next to each other in their tent.

"I know one, definitely. Rhodiola, cornflower, and wormwood?" Pino hazards, not very convinced.

"No, wrong. Two out of three, try again. You have another five seconds."

"Rhodiola, cornflower, and mint from Turkestan."

Tommaso makes the throat-cutting gesture that means he's going to have to pay penance.

"Two out of three, wrong again, It's chifir."

"Damned chifir." Pino throws himself backwards. "Isn't it illegal?"

"Yes, some ex-prisoners we met a few days ago had it," says Tommaso. "They used to use it in the north, in the Komi region. I swapped it for jasmine tea."

---

16 Of course, in Russian.

Nika holds out her hand to try the essence.

"Oh no, not you!" says Pilar, stopping her. "I'm sorry, you can't inhale this yet."

"Why not? We are closed in here with you. We are already breathing it in passively."

"Exactly, if they catch us we're in trouble," Pilar answers.

"So why do you do it?"

Pino opens his mouth for a giant yawn and lets out another series of coughs. He rubs his eyes, and shaking his head confirms the essence is truly psychotropic.

Tasia looks at Tommaso, and then Pilar. "Do you really want to know?" she says to Nika, who nods. "Well, we do it to feel light like smoke, evanescent like a perfume."

A little offended, Nika takes Askalu's hand and tugs at him, wanting him to leave the tent.

"Forget about them," he says, following her out to where Bielka is waiting for them. They walk for a hundred meters before sitting down on a tree stump. "Do you want to play something?"

"Something old or new?"

"Old, a song my father plays a lot when he's angry and wants to calm down. 'Hard Sun' by Eddie Vedder. Would you like to improvise on percussion?"

Nika and Askalu open the musician app and synchronize their phones.

"Would you prefer to compose or shall we use the icons?" she asks, as she tunes her guitar.

"The icons'll be fine," answers Askalu. A set of virtual percussion instruments is projected in front of their eyes.

The heliotrons on Nika's arm light up in lots of bright colors, almost as if they are reflecting her mood. Around her is a cloud of energy she is emitting; Askalu has the impression it is anger.

When she finishes, Nika breathes through her nostrils, as if to smell what she has just produced.

"Do you feel better now?"

"Yes, and no, I wish we were already in Rome."

Askalu is shocked. Wishing for the cancellation of space is like swearing for the Pulldogs, a serious insult towards their favorite dimension. He has frequently heard the adults swearing and cursing time, but he has rarely heard anyone angry at space.

"Do you know Europe?" he asks, trying to lighten the tension.

"No, this will be the first time. One day I would like to go to Versailles."

"Ah, to see the palace of the emperors of France?"

"No, to go to the Osmothèque and sniff all the containers where the original perfumes were preserved."

"Oh, are you sure?" asks Askalu, wrinkling his nose in disgust. "Those essences will be so old they will only smell of death."

"They won't stink, they are authentic aromas, not artificial ones like the ones in the sprays or colored phials they sell in shops."

"Uhm...I wouldn't like to sniff the glands of a deer. It would be beastly."

"No, no, it must be incredible. It must be like traveling back in time and smelling the stink of a mammoth, or a ginko biloba."

Bielka growls quietly, then a sound in the trees puts them on alert.

Two yellow eyes glitter in the moonlight. A little below them, another two, smaller, do the same.

"It's a wolverine with her cub. How sweet...."

The animals are stout with long necks and short, strong legs.

"Aren't they dangerous?" Askalu asks. He has never seen any before.

"If they are hungry they might even attack moose or reindeer. Here, throw them these," Nika says, giving him a handful of dried fruit. With a jump, the animals pounce on the food which they chew greedily before disappearing back into the forest.

"Talking of mother and child, I wanted to ask you something...."

Askalu nods.

"Today, why did you tell Ivan Rafabel was your mum?"

"Because I wish she was," he says. He takes a phial from his belt and hands it to Nika. It contains an infusion of cherries collected during the day.

"Thank you. But she isn't your real mother."

"So? For me she is."

Nika doesn't understand. She drinks the infusion and strums a few chords. Silvia told her Rafabel has no children, and clearly there is no way she can be Askalu's mother.

"But just wanting her to be your mother doesn't mean she is." There is naiveness and clumsiness in her words and his mood darkens. He doesn't understand what Nika is trying to demonstrate.

"It's like she has adopted me," he says after a few moments, almost whispering.

She starts playing again, and that seems to be it, but after the last chord Nika can't help herself.

"Who is your real mother?" she says, spontaneous and direct like a bolt out of the blue.

Askalu turns his back on her and doesn't answer. She puts a hand on his shoulder.

"My mother is dead. She's gone," he says, shaking off her hand.

Nika stops, ashamed. She shrinks into herself. Lowers her gaze. "I'm sorry, I didn't know...."

"Now you do." Then, acting on an impulse, he finds the words to bite back. "Anyway, Alan isn't your real father either."

"What are you saying?"

"What I said. Alan isn't your father."

"You're lying. I'm sorry if I hurt you just now. I just wanted to know why you wanted Rafabel to be your mother so badly...."

He turns and shoots her an angry look.

"It's not a lie. That bracelet wasn't Rafabel's. I threw her bracelet into the sea in Africa, from the deck of the Green Ark."

Nika looks at the object circling her wrist. Now it is her turn to say nothing.

"Rafabel traveled a really long way to bring you those gifts. She wanted to tell you who they were from but your parents were too scared."

"Scared of what?"

"Those gifts are a kind of legacy from your father, your real father."

She wrinkles her forehead. "Legacy? What are you talking about? What's it got to do with you?"

"Legacy because your father is dead. I was there when it happened and I knew him."

"Who was it? My parents' friend? The man in the video they didn't want me to see?"

Askalu nods with his chin. She doesn't know what to say. She leans her guitar against the stump and stands. She has to walk, to move, like she has always seen her father do. Her father…who is no longer who she thought he was, no longer a person, no longer a precise identity.

"That's why they were so strange. But why didn't they say anything to me?"

Askalu lifts the phial with the infusion.

"Cherries. We are still green," he says with all the maturity of his eleven years.

Nika doesn't feel angry, neither with Silvia nor Alan. She can feel a pain; perhaps it is more a regret for something that has happened and she could do nothing about, but at the same time she understands their motives. Her pain is in reality closer to discomfort, bitterness, realizing that truth has the same consistency as the clouds, you can see what you want in them but you can't touch anything, nothing is more than a fleeting shape.

Askalu shuffles closer to her. "Don't be sad, Nika, you have a mum and a dad who love you. Now you know you had another father too, he was an incredible man. I saw what he did in Sangha, my village. If all these people are going to Rome in remembrance, you should be proud to be his daughter."

"But I don't even know who Nicolas Tomei is!" Her voice is deep and high at the same time.

"At the end of this journey you'll find out."

He takes the phial back from Nika. "Do you want to laugh? Let me tell you something about my father."

CHAPTER SIXTY-TWO

Greenternet

At dawn, the forest is spread with a quietness and devoid of sun. The Pulldogs proceed along the walkersway through the undergrowth of rotting logs, blueberry bushes, colonies of blackened fungi and an ocean of leaves – millions of leaves – like froth between the tree roots.

Today the dendrophora is talking about the economy. With long strides, she is asking her pupils to imagine the market as a huge pollination process.

"In the plant world," Shan is saying, "no one does anything for nothing. Whoever takes something or requests a service has to do it in exchange for something in order to construct a system of reciprocal advantages. It's similar to human economy but without speculation, where people and things are exploited for personal gain. For example, insects pay with their work, whereas plants use their nectar as a kind of universal currency, a sweet energetic substance which animals like so very much. Trees use it in exchange for the transportation of pollen by insects, and use their fruit to convince animals to take their seeds far from the mother plant."

Nika hasn't slept very well. At breakfast she avoided talking with her parents. She has decided to wait before telling them what she has discovered. The turmoil of thoughts in her head stops her from paying proper attention to the dendrophora's lesson and its strange metaphor of insects as buyers, pollen and nectar as merchandise, plants as sellers, and even colors and perfumes used as advertising.

All she can think about is the perfume Rafabel brought her. The temptation to sniff it was too strong last night, and in the private space of her tent she had unscrewed the lid and let herself absorb the essence. What she experienced this time was very different from the last time. Aromas are like places in memories, they are mute if you don't know them, they whisper quietly if noted in passing, but they reveal a whole past if they are connected with something you have lived through.

Nika makes an effort to listen to what the dendrophora is saying, but the words are still only getting through to her like weak, blurry signals.

"Whether it's apples, cherries, or apricots, the flesh of the fruit serves two purposes: to protect the seeds until they are ripe, and as recompense for the couriers dealing with their transport."

Nika's thoughts are invaded: a mysterious golden fragrance that tantalized her nostrils and crept along her olfactory nerve unleashing completely new associations of ideas/experiences/sensations. It had felt like she was traveling across thousands of kilometers, living for hundreds of years, as if the phial had imprisoned a genie capable of unleashing the flavor of unexplored territories and evoking the odor of time that has passed, of physical and mental transformations. The fragrance changed gradually – from woody to spiced, from balsamic to minted – it burned her nose's mucous membrane it was so intense with an encompassing character, citrusy and fermented, earthy and floral, with notes of resin and honey, tuber and salt. It imposed itself, immediately becoming a memory. This smartfume is a more insistent amalgam than any other memory or fantasy Nika has ever had in her short life, comparable to an effervescent novel to be inhaled, in parts bitter and sour, in others sweet and flavorsome.

The explosive synthesis of smells took her over sandy dunes, to rocky cliff faces, following animal tracks, and experiencing the effects of stupefying inhalations, wrapped in invisible webs, seeing bizarre insects and mellow birds. She felt as if she had seen whole starry nights, watched raging torrents, taken breaks alongside dusty roads, run across fields, crossed oily seas, rested below dry trees, descended steep hairpin bends, been approached by funny people, spent time alongside wise friends and sinister presences...what was living in that cloud? Her father had poured his essence into an olfactory memory, an intimate and personal dimension as intense as the universal immediacy of a picture, evanescent and pervasive as a person's individual odor.

A persistent joy descends from the reality of the senses – Nika wouldn't know how to express it in words – and that perfume captures its purest essence.

Finally, Nika returns to listening with her ears, after listening with the memory of the odors. The words of Shan are no longer coming from in front of her; the group has slowed down without her noticing.

"Tomorrow we won't be having lessons except for an exercise in arborography at the end of the day."

At this the kids all scatter out along the path.

The next day, the Pulldogs reach Bashkiria and the taiga gives way to the steppe. The slopes become barren, and beneath the low, sparse covering of the vegetation they catch glimpses of slate-colored rock covered with lichens, the famous černozëm, the black land, extending up to the southern Urals.

As the woodland coverage diminishes, the trunks get smaller, the wide canopies vanish, and where previously the Pulldogs were walking in the shade of the foliage, they are now exposed to a shy summer sun veiled by clouds. Over the past few days, other communities of wanderers have joined them. On Global Walker some groups and single walkers have signed up to the idea of going to Rome to celebrate the figure of Nicolas Tomei. In the midst of this, Nika has witnessed a number of painful separations, and for the first time she has noticed her father attempt to convince some people to stay with them, unlike his usual 'live and let leave' philosophy.

"We will follow the example of other caravans who have come before us," a bearded blond with his wife and two children said to him. With them were three young couples who had recently got married in the waters of the River Irkutsk.

"Who were these caravans? Do you know them?"

"We have never shared a route, but on Global Walker I saw they have left the forests and settled alongside ancient abandoned roads."

"They have become stationary?"

The man had lowered his head. Then his clear eyes lifted and they were filled with pride.

"They miss the firmness of the asphalt, they are nostalgic for a lasting identity, they don't entirely understand the sense of freedom."

"How have they reorganized?"

"The indispensable minimum. Gardening, off-grid energy and fabtotums, but they have stopped moving. Some people call them the guardians of the road oases."

"And do you want to be like them? Living in service stations?"

"I was a medic before ecolution, I cared for patients in the villages around Perm," he said, turning to look behind him to show he was speaking for the others too. "We don't want to be on the roads too much. We prefer collective farms."

"The Kolchoz?"

"My grandfather lived on one in the Republic of Komi."

"I respect your choice. Our paths separate here," said Alan in a solemn voice, using the formula coined for such occasions, "but everything the paths divide, the paths bring back together."

That same evening, while they are preparing the nest, Nika witnesses the opposite. A solar plexus appears from the woodland, moving slowly. He is limping, and when he is close enough they can see his skin is sun burned and his head is shaved and shining with heliotrons.

A group of people are beginning to gather in a ring around him, the usual behavior when people want to see something strange while still maintaining a safe distance.

Miriam approaches him to find out where he has come from.

"Ulyanovsk...Toliatti...Samara..." the man struggles to answer. Every now and then, he closes his eyes tightly as if something is blinding him despite the fact the sun has set.

Suddenly Miriam recognizes him. He is one of the first plexuses that she herself created, years ago. "Are you feeling all right, Igor? It's me, Miriam, don't you recognize me?"

Igor used to be a builder working for a company contracted to build a hospital in an isolated valley in the Novosibirsk region. Unfortunately, once the steel frame was finished, a fire destroyed the building site and everything else within a radius of forty kilometers.

"Miriam...so much time. It's the old pesticides, containment structures have crumbled. The antiparasitic...gone, washed away with snow and rain. People still use them, in small quantities...in gardens, at home. They don't understand the risks."

His breathing is labored.

"The nanites die on contact with oxygen, they were designed to survive inside organisms. Do you mean pesticides did this to you? Has your immune system been updated?"

Igor had studied agricultural engineering, but had had to take on a building job. He loved greenhouses, one of the rare cases where resilience was the fruit of a collaboration between humans and plants, and he told Miriam he had built one in his *dacha* to grow cucumbers, potatoes, and tomatoes. He dreamed of growing an avocado tree one day, a fruit he had tasted once in Turkey during a vacation on the Black Sea.

"I don't know how it happened, theft, corruption, or fires... hard to understand, but the pesticides united with the nanites... using them like vectors...to enter other organisms."

When his greenhouse was destroyed by a summer storm, Igor decided to give up on everyone and everything. Miriam found him by chance; he was trying to kill himself, like many people in this area, drowning in alcohol in the middle of the road. She pulled him back together, gave him a phial of heliotrons to swallow and a good reason to carry on living: the hospital wasn't the only place where people could be looked after, a greenhouse wasn't the only way to feed himself.

For Igor, ecolution became a kind of mystic engineering. In the empty space left by communism and by the consumerism of turbo-capitalism, there was space for something new, something to reconcile the desire for individual freedom with a more aware lifestyle.

Igor's condition is critical. People start giving advice.

"Bring water!"

"Make him lie down!"

"Cover his skin!"

The dendrophora, on the other hand, leaves the path of the walkersway and takes a few steps into the forest. First, she rests her hands on the trunks of some beech trees as if introducing herself to them, then she chooses one and caresses its branches all along their length to the leaves. Then she takes a thin needle on the end of a USB cable attached to her phone and slips it into the tree.

Nika watches Shan and remembers the lesson on olfactory messages.

"From the roots to the leaves, plants use volatile molecules to receive data and communicate with each other and insects. The receptors are like lots of different locks and the odors are the keys: each lock opens on contact with the right key, and this puts the mechanism that produces the olfactory information in motion."

After putting the needle into the trunk, she studies the DendroScan App on the screen.

"You are right," she says loudly after reading the results. "The beech is producing enormous quantities of methyl jasmonate, a molecule it emits under high stress."

"How long have you been in this area, Igor?" Miriam asks.

"Two months."

"The direct toxicity is not as dangerous as prolonged exposure and accumulation," the dendrophora says. "Pesticides enter the organism and spread through the liver, nervous tissue, and the other organs. That's why Igor is in this state."

"What're we going to do?" Miriam asks Alan.

"We have to warn the authorities and run, so they don't blame us," says Ivan. "I'm going to flag the danger on Global Walker."

Alan nods and adds, "We'll take a diversion towards the south. Shan, can you map up to where the plants are emitting this molecule?"

She nods.

"We will walk around the area, travel parallel to the Volga, then we will cut across towards Rostov-on-Don. Dikran, can you check Global Walker to see if there are other similar reports down there? Ivan, what do you say about contacting Kirill? We will have to change our rendezvous with the Green Ark."

"We can't just leave," says Rafabel. "In Africa we came across a similar problem, a virus engineered to make people delirious. I know someone who might be able to help us."

Alan's worried face has the same expression as when he uses the guitortoise as a weapon rather than a musical instrument.

"All right, if your friends find a solution, they can spread it over the mesh net."

A minute later, Dikran confirms his suspicions. He takes Alan to one side and shows him the screen. "There are alarms all along the Volga up to Saratov. We'll have to be careful," he says in a low voice, "to avoid the risk of contamination."

Dikran rubs his hand over his forehead and looks at his damp palm.

"Are you thinking about sponges?"

"Perhaps.... It's a drastic measure, but necessary in a case like this."

"How long will we have to resist?"

"The contaminated area is large," Shan says, "around six hundred kilometers." She points to a red zone on her phone's screen. "We can get out of the area in three days if we run as fast as we can."

"How long will it take to compose the nanites?"

"If we start work immediately," Dikran says, "they'll be ready tomorrow."

"Great. Ivan will be able to help us. In the meantime, we'll all have to wear masks."

The next morning Igor's condition has deteriorated so much he asks to be taken to water. Miriam knows the significance of this request.

The Pulldogs leave the walkersway path and go into the wooded hills. They won't be able to eat seasonal produce and they will have to test everything they collect before eating it, lengthen their fasting, and at the same time raise their metabolism to be able to run as fast as they can. At their first stop they all queue up in front of Dikran and Ivan. In the bottle they pass from mouth to mouth there's a nanite-enriched liquid.

"Nice!" says Nika, drinking the solution and licking her lips.

"What's in it?" asks Askalu.

"*Kompot*," Ivan says, "wild fruit juice, pears, sloes, apricots, and dried hawthorn berries; we had already made reserves for the village. Now we have also added special nanites…. Over the next few days we will be like sponges. Sponges usually stay motionless on the sea floor and absorb a one hundred liters of water per day to feed themselves. If the water becomes murky, they emit a series of electrical impulses to their cells to interrupt feeding, and therefore the absorption of damaging particles that might obstruct the sponge's pores. In the same way, the nanites we have just taken will make our skin impervious to external agents, not forever, but for enough time to escape the toxicity of the area."

After an hour of running, they reach an artificial lake. The warm water feeding it comes from a rusty pipe resting on a line of Kavkaz Service truck tires. The end of the pipe disappears into a hole dug in the ground; it's probable the flow comes from an abandoned industrial plant and ends up here after having passed through who knows how many *izbas* scattered in the area.

The solar plexus is laid out gently on the grassy ground. The call of the water is very strong. The adults keep the kids back; only after the dendrophora has sampled the liquid and given her permission can they all start filling their water pouches.

"*Podsolnukh,*"[17] mutters Igor. He unclips his bracelet and gives it to Miriam.

"Of course," she says, accepting the task implicit in the gesture. "I'll find a good place along the walkersway."

An ancient tradition typical of Russian spirituality consisted of building *obydennye cerkvi*, small churches erected in one day thanks to the teamwork of a community. They were the perfect symbol of cooperation enabling rural villages to survive, a concrete example, non-mystical, of a 'miracle': a church built in the present, where before there were only trees, an anonymous place transformed into a space to get to know each other again, without supernatural forces or divine intervention, but thanks to collective desires and the commitment of everybody. There were more than six hundred chapels of this kind that had been knocked down in recent years and transformed into off-grid recharging sunflowers scattered along the walkersways, the *podsolnukh* Igor was speaking of.

They help the man into the lake, a few people go in up to their knees with him. Every now and then, Igor looks imploringly behind him but without showing any signs of stopping what he is doing.

Miriam looks worriedly at the man and then back at the Pulldogs. In an attack of desperation, she asks parents to cover their children's eyes.

"Let's at least spare them the end," she says bitterly. With a ghostly frown, she gets into the pool too. The more Igor moves in, the more the heliotrons slow down: his skin goes from mauve to indigo until they lose their color and the plexus returns to looking like a biological human. At the end, with the water up to his chin, Igor lifts an arm and points upwards.

---

17 Sunflower, in Russian.

His last upload. All of his energy is about to be transferred to the ergonet cloud.

"Live by fire, die by water. Live by fire, die by water. Live by fire, die by water." Igor repeats the phrase like a mantra and continues his descent until he disappears.

"Live by fire, die by water," Miriam echoes. "Now they can see."

Nobody does anything, nobody moves.

The shine of the heliotrons is extinguished by the dark of the pool, and the solar plexus ends his path under a meter of water. Dissolving in that chaotic pool, Igor has given a meaning to his world.

CHAPTER SIXTY-THREE

Fathers and Fathers

"Mummy, I'm tired. Can we stop?"

"Not yet."

Along the open stretches, the Pulldogs spread out, on the dirt tracks or on the country roads they move closer together to become a single file. Panting, synchronized legs, and eyes staring straight ahead, they move as one: the young in the middle, the adults at the front and the back.

"Daddy, I can't run anymore. You're going too fast," Nika moans.

"Don't talk. Don't think. The brain is lazy."

The June sun is testing them more than the actual escape from the red zone. Sweat-stained clothes are a rarity, a direct consequence of the nanites' frenzied action: the answer to the urgency for maximum energy is secreted in a continuous stream. It liquefies on their burning skin and evaporates in front of their eyes.

When they come to farms, isolated houses, family yurts, and camps, they warn the people of the danger without stopping for longer than a moment. The journey continues at a sustained pace. They run far from the noisy sides of the motorways, through almost-empty villages overrun by the stronger plants, over hills in flower, and across sleepy plains beneath the silent trails left by airplanes high above.

At dusk on the third day a vibration echoes through the nest. It's Alan playing his Tibetan trumpet. The instrument's long notes, deep and sad, alert them to a general assembly. Everybody looks out of their tents.

"We'll be leaving soon, in the night," he says from the highest branch of an oak tree. "Shan has confirmed the contaminated area is behind us.

We can head towards Rostov-on-Don, and if we hurry we will be able to board the Green Ark within three days."

Nika goes to Silvia and Rafabel, looking for produce.

She sees they have collected a basket full of forest fruits. There are also acorns, chestnuts, and sunflower seeds.

"Can I help you?"

"Of course," her mother says. "If you find something, mushrooms or fruit, give them to me so I can check them on the DendroScan to see if there's any residual contamination."

Nika has chosen this moment carefully. She has waited too long without doing anything about it. Igor's end has disturbed her dreams and made the days of running sad. Now their pace has slowed and her thoughts have gone back to normal, this is the perfect occasion.

"I get it now why Daddy was agitated when you gave me my birthday gifts."

Silvia and Rafabel look at each other in surprise.

"Really? Maybe you should talk to him about it," says Rafabel, trying to deflect the blow, but Silvia realizes the time for innocent lies is over, the time for uncomfortable truths between mother and daughter is here, how painful it is going to be remains to be seen.

With a knowing smile, Nika twists the bracelet around on her wrist.

"This object.... I'm not angry you didn't tell me anything. But I want to know now too."

"Who told you?" Silvia asks. At this point all three of them have placed their cards on the table.

"Askalu. The other day we were arguing because he kept saying Rafabel was his mother, and I didn't know his real mother was...well, I was stupid, and then he got his own back by saying that Alan wasn't even my real father either."

"What else did that naughty bigmouth tell you?" Rafabel adds, trying to lighten the tension.

"Everything. He told me about Nicolas, the gifts, your attempts to 'protect' me," she says, wiggling her fingers by her ears to put the word 'protect' in air quotes.

"What would you like to know?" Silvia has stopped and rests her basket on the ground.

"I want to know who my father was. I mean my other father."

Silvia and Rafabel share a perplexed look.

"I don't think Daddy would be very objective about Nicolas, do you?"

"Right." Silvia hugs her daughter. There's a flash of understanding between them, which means she has entered, at least partially, the world of the grown-ups.

"I mean," the little girl goes on, "you both knew him better than anyone else. Except for his mother and father."

Both women burst out laughing at this.

"Nicolas's family didn't know what to do with him," says Rafabel. "At least, they wanted him to be very different to how he really was."

"Does that mean I won't have an extra two grandparents when we get to Rome?" Nika asks.

"Aha," says Silvia, opening her arms rhetorically, "my daughter feels she doesn't have enough relatives!"

"I don't know, do I?"

"Believe me," Silvia says bitterly, "if you ever have the misfortune to meet Pietro Tomei, you will wish you had been born without a grandfather!"

"Well, then? Who's going to start?"

"I'll start as we walk back to the nest. Ten minutes won't be enough though, but I will start by telling you something about his childhood, seeing as I knew him best when we were kids in Rome."

All three of them start walking towards the nest carrying the full baskets.

"I remember in middle school Nicolas had so many nervous tics, he kept fiddling his fringe into place, which was strange because his mother cut his hair really short once a month; he used to look at his nails like girls do, only he had no varnish on his. He wore checked shirts and he wouldn't stop touching the collars, as if they were never straight enough. His world was full of awkward fantasies, weird and sometimes hysterical to other people's eyes. For example, there was a period when he would self-diagnose himself with all sorts of illnesses. When he was twelve, he went to the doctor more than any of his friends in the neighborhood. He would get rashes and outbreaks, normal for his age, but he would cover his face with plasters. He used to phone me for the strangest things; he had a little pain in his side, his elbow, neck, and he was filled with anxiety. If his thumb got itchy he would panic."

"Were they real or did he make them up?"

"Some things were real, others he exaggerated. He used to make a drama out of everything, not least because he was under the obsessive influence of his mother's and Pietro's protective wing. Even the smallest incident, told by him, became a tragedy he had escaped by a miracle. Incredible things were always happening to him; one year his moped managed to get a punctured wheel six times. He was habitually late for school and every time he would invent a different excuse. He never had money with him, even though his parents never left him short. I gradually began to weigh up everything he said because I found it hard to believe him, he invented silly stories, maybe some of it was true...but most of them were infantile techniques for attracting attention. Some kids made fun of him, others ridiculed his mishaps."

"And what did you think?"

"I was his friend. His family was difficult and that's where his problems came from. Every now and then, we would go in a group to Trastevere. He would take us to the Botanical Gardens. His father forced him to go to learn to recognize the perfumes, especially when it was most difficult, after storms brought by the sirocco wind, but Nicolas preferred to sift through the fallen leaves looking for something precious. He was looking for a magical object, a personal amulet he could use for strength and courage, something ancient and powerful to use in his fight against bad luck."

When they get to the nest, Rafabel links arms with Nika. Bielka runs to her excitedly after a separation she considers far too long.

"If you come with me to my tent," Rafabel says, "I'll tell you something special about Nicolas."

Nika says goodbye to her mother and goes with Rafabel. After five meters, she and Bielka climb up the wooden ladder of the treehouse Rafabel has composed. Nika settles down on a platform three meters up the tree. Bielka enjoys the view too, paws hanging over the edge.

Rafabel offers Nika a crisp piece of larch bark and an infusion of birch she has prepared using a phial from her bandolier.

"Nicolas wanted to send you a letter for your sixth birthday. We were in India, in Auroville for the *Holi* festival. He was happy, he felt fulfilled and wanted to share this emotion with you. He dictated a letter to me, then he had second thoughts. He worried maybe you were too small, you wouldn't understand, and anyway it would have harmed you to hear his words. His presence, an apparition out of the blue, would have confused you and made your life more difficult. So he asked me to throw the letter away."

"But you didn't...."

"No, I think confusion and difficulty have a purpose. Sometimes men don't understand the subtleties," she says, wrinkling her nose ironically. "So I pretended to throw it away but kept it."

From a compartment in her bandolier, Rafabel pulls out a tatty piece of paper and hands it to Nika who starts reading. "It would be nice to start this letter by telling you who I am, seeing as we haven't met. But who I am is a difficult concept to explain. Anyway, here we go. I see myself more as a territory than a person, and I think you are probably already laughing. But bear with me, for me, 'who I am' would be the roads I took up until a certain age, and then the paths I have walked along for the rest of my life. Perhaps I would have done better not to start this letter with 'who I am', I mean, perhaps telling you where I have been would be the best way to tell you who I am. This is because we are not only the product of a moment, or a period, or an entire life, but also of the biological and social forces that shape our bodies, guide our perceptions and influence our thoughts...."

"You don't have to read it all out loud. He wrote it so he could spend some time alone with you."

Rafabel has tears in her eyes. She too has become Nicolas. She too, step after step, has shared the trajectories of his transformation. Something hurts her because – unlike him, and especially without him – she will go back to feeling like a neutral territory, nothing wonderful or amazing like when he was by her side. Yes, perhaps he hadn't loved her as much as he had his various projects, perhaps he had considered her as an excellent traveling companion, a loyal and passionate confidant who would follow him anywhere, but love.... Not even on his deathbed had love convinced him to fight back, to stay alive to give her as much as she had given him. For his part, that was the greatest degree of love Nicolas had been able to show a human being, including, perhaps, his own daughter.

Her voice breaking with emotion, Nika rereads the last passage aloud.

"The bracelet has witnessed all of my travels, it has followed me everywhere I have ever been, and, step after step, I have become as one with it. Happy birthday and *bon voyage*, Veronika!"

Nika gives the piece of paper back to Rafabel. She doesn't understand everything Nicolas was trying to tell her, but the letter has left her with a pleasant feeling like when, looking at a view, she grasps its extent, profoundness, and splendor without being able to pick out any details in particular.

"No. Keep it. It's yours. I should have given it to you together with the bracelet...."

"Thank you," she says, but she still hesitates. "So, if I have done the calculations right, you're a kind of stepmother for me, and Askalu and me are kind of stepbrother and stepsister, right?"

Rafabel thinks about this.

"You know, I hadn't really thought about these relationships, we have always been part of an extended family, ever since we all lived together in Serra Spino."

"What was it like living in a city, always in the same place?"

"Well, we had neither the advantages nor the disadvantages of the nanites. We had to find food to eat at least twice a day.... We had to wash to remove dirt and pollution from our clothes and our skin. We had to earn a living because you needed money to make anything work. But it was comfortable...yes, comfortable is the best word. If you accepted the rules of consumerism, there were no problems. Maybe I'm saying things you don't understand, but I can guarantee you'll have your first culture shock in Rome and you will meet people very different from us."

Nika puts on a questioning expression and brings the conversation back to a level closer to her. "And what were the two of them like?"

"Silvia has already told you about that, hasn't she?"

"Yes, but..." Nika looks around her as she searches for the right words, "...you have a different viewpoint. Like from high up in a tree."

"I suppose you want to know about Alan and Nicolas."

She nods gently as she nibbles on her piece of bark.

"Alan has always been there in the life of our community, but he wants to be free, without having to report back to anyone. I think that he might be at the point of not respecting other people's rules and wanting to make his own. He has always hated society, and at the same time he has never been able to manage without his family, his group. It might seem like a paradox, but that's the way he is."

"And Nicolas?"

"From the first day I met him it always felt like he had to rid himself of guilt. Perhaps that's why he was always absent, his head was always somewhere else and his heart cracked by who knows what anxieties.... He might have seemed self-centered to many people, because he wanted to reach objectives that would convince him he had finally left his feelings of inadequacy behind...feelings he had been dragging around with him since his childhood."

Nika is finding this difficult to understand, so Rafabel tries hard to be clearer.

"Nicolas was self-centered because he was dissatisfied with himself. First he had to lose weight, then free himself from the influence of his family, then save the Pulldogs, and finally free the world from its dependence on food. Underlying what seemed like egoism there were many dreams...only those who understood this managed to love him."

She stops for a moment to gather herself together.

"If you had to compare them, Nicolas was egoistic as a person but he had many ideals, whereas Alan cares for his family, and doesn't give a fig for all the rest."

CHAPTER SIXTY-FOUR

Rockambulance

The Pulldogs are given a warm welcome by the crew of the Green Ark. Kirill is happy to see them again but also worried.

"What's the matter, my friend?" Ivan asks after a bear hug. "Has the annual budget been renewed? I was only on land for three months...."

Ivan knows the problems connected to the project well, the attendant bureaucracy needs constant attention and caution, especially in an endeavor like theirs, always at risk of being suspended, if not disbanded by request of some international moratorium against nanotechnology.

"No, that's all in hand," answers Kirill. "We won't have any financing problems from above for the next eighteen to twenty-four months. The approval of our budget arrived last week.... The real problem is it's getting more and more difficult to pass unobserved and not have problems *from below*."

This conversation barely reaches Nika as she – guided by Silvia and Rafabel along the paths on the vessel – is amazed by the geodesic domes. Since the last time the Pulldogs were on the Green Ark, the keel has grown and thickened like the trunk of a tree, the aquatic roots have lengthened, becoming a cumbersome trail hundreds of meters long, the phosphorescence of the tree canopies is an irresistible call for the numerous species of migrating flying creatures and for those that decide to nest in their branches. Unfortunately, the same call attracts curious people on the lookout for bizarre content to post on the web, and the harbor authorities of the Mediterranean who detect them kilometers away can't wait to pass them under a nano-detector.

"The old and new *peerates* know us and warn us about patrol boats," continues Kirill, "but we are seeing more and more tourist drones and businessmen looking for investments on behalf of some sheikh or business group. At the moment, the most urgent thing is not to be blocked from the Dardanelles. Years ago, they stopped us for two weeks, confiscated a rare specimen of *Wollemia nobilis*, and fined us ten thousand euros. Things are different in Turkey now, the country is divided about nanites, but the authorities can invent any excuse they like if they want to stop us."

"So, why don't we invite them on board?" says Alan, surprising them all.

"What do you mean?"

"Let's impress them with our presence. We can set up an interactive performance *for* the trees, *with* the trees as absolute protagonists. We can organize a distributed computation fundraiser with the patronage of the United Nations. Our friends at Prinkipo on Büyükada can give us a hand with the hardware. The trees will not only be our musicians, they will also create a breathtaking backdrop."

"Are you sure that'll work?"

"I've already done it. A floral-musical in the Nong Nooch botanical gardens in Chonburi, Thailand. Local artists helped us apply a network of sensors to the leaves and roots of some Japanese Sofora; we used a special device to pick up the signals of the root-leaf lymphatic movement captured by the sensors, and then we recorded the electrical resistance of the tissues."

"You gave the plants an ECG?"

"More or less. Then we passed the sequence through a sampler which transformed it into digital signals, and therefore notes. We will use rock to save the trees. I'd challenge any authority to go against an idea like this." Alan looks to Ivan and Kirill for support. "I'd go as far as to say we have a name too, 'Rockambulance: plant beat music'. It sounds like something Nicolas would have thought of, doesn't it?"

The two older men seem to be convinced.

"Oh, why not, let's do it. We'll be there in two days. How long do we need to organize the rest?"

"A week should be enough."

In the meantime, Nika slips away, she says 'bye to her dad and follows Bielka, who is running madly through the flora of the temperate dome.

They choose the narrowest point of the Çanakkale promontory. The night is clear, visibility is excellent and from both sides of the Dardanelles thousands of people have gathered to watch the tree concert, from the Kilitbahir fortress on the Çanakkale seafront right up to the surrounding hills. Applause, long whistles, shouts of enthusiasm, and infinite inhalations of party essences fill the air as they wait for the show to start.

After the speeches from the local authorities, live connections with Istanbul and the Turkish seat of the United Nations expressing their thanks, something begins to move in the dark on the Green Ark dressed up by Esin Demir, Doctor Çakmak, and the Prinkipo team.

From a distance, sparks and glows can be seen; they can make out creaking sucking sounds, whispers and popping noises. Then, suddenly, there are flashing sequences of colors like mosaics of leaves, grid-like patterns along the branches and tree-like shapes that disappear and reappear in new shapes and outlines, their intensity mutating. The shades of colors are similar to a musical scale, rhythmic fugues and melodic emulsions.

Dressed in black, wearing a green beret, Alan stands out in the shadows of the geodesic domes as he picks at the guitortoise and composes the notes, literally extracting them from the tree branches, dipping into a reservoir of vibrations rising from the roots and descending from the leaves, all then filtered by the trunks connected to software harmonizing this phytophonia.

Pietro is playing bass, Dima, one of the crew technicians, is on the mixing desk, and Askalu has been promoted to drummer. Samples of gongs, rustling choruses, and lymphatic rhythms mingle with sounds creating the soundtrack of a reserved, unreal world awash with non-human timbres and frequencies. The combinations of images, sounds, and colors transform the concert into an iridescent synesthetic experience: a fan of sounds runs and spreads like a fluid in relief, it never breaks nor separates; on the contrary, it dissolves in simultaneous parallels as the pulverized images fade into volatile particles and granular spores, bewitching the audience.

The visual unit is the leaf. The rhythmic unit is the flower.

The root is the bass line. The branch, a note. The trunk, a chord.

The group of domes form an immense sound texture backdrop. From this chaotic mixture, fragments of a melody everybody knows begin to emerge. An old song from the previous century. The vibrations increase and the human beings respond to the vegetable triumph. Alan is no longer plucking his instrument, he is holding and playing it like a note-spitting guitar producing 'Kashmir' by Led Zeppelin.

Hundreds of drones in geostationary flight are filming the performance and broadcasting live, worldwide. The public are no longer making any noise, captured by the exceptional nature of this mysterious phenomenon. The media shack atmosphere of the Rockambulance vanishes.

The notes travel in long metallic modulations, with no vibrato, then they change with dramatic effect produced by an unknown musical instrument, as arboreal phytophonia is unknown. It's something that goes beyond music, art, performance, and show: it is the expression of an emerging community.

The rhizomusicality of the floating forest reverberates the emotions of the audience. Perhaps they were expecting banal plant rustlings, the snap of breaking twigs, the tinkling of viscous liquids and the crunching of piles of trampled leaves. Instead, what they got was the trees, like huge parabolic aerials capturing every energy variation in the environment, modifying sound emissions and tuning in to the general mood. An explosion of bass notes, timbres and drumming lights up every plant.

Alan roars at the end of the riff. Pietro's bass follows and Askalu's drumming accompanies them along a forest of cacti crossed by a stream that ends in a waterfall beyond the ark. Now everybody is dancing, jumping, shaking off negativity, and shouting out their surprise, getting rid of their tension.

As the performance slows, Alan's voice wanes and shakes with the agitated rustling of the leaves. The music empties thoughts, the rhythmic flashes weaken, but the vibrations continue.

The hashtag #rockambulance will be the most commented for the following eight hours.

"How did it go?" Alan asks, drying the sweat from his forehead.

"It was great. From now on, we can use our fame as a free pass for Turkey," Ivan says, satisfied.

He and Kirill are inhaling an essence to celebrate the raising of a quantity of distributed computing sufficient to maintain the Green Ark for another two years without extra financial help. Silvia and Nika congratulate him, but Askalu is the one who notices something strange. "Your instrument is breaking," says the boy, pointing at where there is a crack in the curve of the shell.

"I hadn't noticed. I must have battered it about too much recently."

Checking over the surface of the guitortoise, Alan finds a crack where a tiny brown bean has got stuck. It is about a millimeter long and has just started to sprout, making the instrument's shell split.

"Will you lend me your nails for a moment?" Alan says to Silvia, handing her the guitortoise. She removes the seed and shows it to him, holding it between her fingers.

"What plant is that?"

"Wait," Nika says, turning on her phone. "It could be an acacia or...." She takes a photo of it and opens the nanomat's database. "Yes, it's kudzu, a Japanese climbing plant. It's used to make drinks and puddings with apples and plums. It is produced using clones identical to the mother plant from the branches that put down new roots."

Alan tips his head and looks at her good-humoredly. "You know everything...."

"It's all thanks to Shan Jao and her botany lessons. Oh, and one more thing, the kudzu is one of the fastest-growing plants in the world. Up to thirty meters a day," she finishes.

Alan's face lights up as if he has had a revelation. With a sly smile, he goes to Kirill. "Do you have any kudzu seeds on the Ark?"

"Of course, in the store, down in the hold."

Alan takes the tiny seed from Silvia and looks at it more closely. "Good, I'm going to need a few hundred."

"May I ask you what for?"

"I've had an idea," he says, looping an arm around Kirill's shoulders.

"Is it dangerous?"

"Not really, but to make it happen, we have to go back to Italy first."

The group goes to the main deck.

"No more than five days and we will leave you in the international waters by Rome."

"What about the Naval Blockade? Is it still standing?"

"Yes, it moves to wherever new routes to Europe are reported. We can get past it but we can't stop. At best, we will be able to get you to the Ventotene Marine Reserve. A dinghy can take you to shore by the Pineta di Ostia."

Kirill leads Alan, who is lost in thought, inside the Ark, rubbing the seed through his fingers like the bead of a rosary.

"It will disintegrate…" says Alan, going down the stairs. "Sooner or later, every concept of borders, even national ones, will disintegrate, as soon as the idea of belonging to a nation is seriously questioned. The nation as a political unit is obsolete."

"I hope you are right. Unfortunately, the world hasn't changed and still punishes the few people with an imagination that others don't recognize as a possible reality."

After going down four ramps of stairs, they turn into a corridor.

"You are right, Kirill. But let me tell you something, three months ago I was going back to the nest after a concert in Kazakhstan with the group. In the middle of the taiga, the border police stopped us and took us to the station where we had to stay the whole night. The day afterwards a civil servant arrived, a woman of about forty with icy eyes, asking us where we came from. I told her our nation wasn't a territory but a path. I told her we lived across countries about which we knew the geographical, social, and historic nature, but not the political one. We were men and women with many different languages and cultures, but no passports."

"I'm guessing she didn't take that well."

"Wait, this is where it gets interesting. I told her if she let us go, she wouldn't be letting foreigners go because we didn't live in any city, we didn't occupy anybody else's space, we lived in places not many people know of, that we lived in a technologically frugal way, rich in experiences and poor in possessions. In the end, I begged her to consider us as birds in transit from one migration stopover to the next, transitory human beings."

"What did she say?" Kirill says, spinning the wheel on a bulkhead door to open it. Behind the seed store is an immense space ten meters high and over two hundred long. Containers are piled up to the ceiling in their thousands. On the shelves, in alphabetical order, is the whole of Earth's plant biodiversity.

"Her name was Anastasia Lipkina, and two months after our first meeting she contacted me to ask how to start a walker community. She was tired of the life she was living."

Kirill pulls at his earlobe. "You were lucky."

"Perhaps. But the seed that sneaked its way into my guitortoise is like the meme that sneaked into Anastasia's mind."

Nika and Askalu run their fingers over the containers as they read the labels excitedly. "It's cold in here," she says.

"It helps keep the seeds in a state of dormancy."

"What, are the seeds dead?" Askalu asks.

"Not really," Kirill explains, "dormancy is like a superpower allowing the seeds to spread not only spatially, but in time too. The seeds that make use of it have an advantage over rival plants with short-lived spores. Dormant seeds can't germinate and they don't want to, at least, not until the conditions are right."

Kirill walks amidst the shelves, looking for the kudzu seeds. "What about you, do you live like a seed or a spore?" he asks Askalu, who is following him. He shrugs, not knowing how to answer.

"Spores are like boys," Nika says. "Seeds are like girls."

Kirill stops by a column of containers and pulls one of them out, opens it, and then shakes it gently in front of all of their eyes.

"Here you are, hundreds of girls for you, Alan."

CHAPTER SIXTY-FIVE

Foreign Body

Five days later, staring at the yellowish line of the horizon, Nika has stomach cramps: not hunger cramps, but from her nervousness about going to Rome for the first time.

She can just about make out the details of the beach, the green canopy of Castel Fusano's pines making an agitated backdrop to the silhouette of the dunes. In front of this, a strip packed with bathers – fuzzy colored dots – fills the space down to the lapping waves; a range of shapes, shiny and square, occupy the rest of the scene.

Either side of the motorboat, almost as if escorting them to the shore, they have been joined by jet-ski and kite surfers performing acrobatic maneuvers and skipping over the waves. As they get closer to the beach, she sees people involved in a variety of seaside activities: people are jogging, doing tai chi at the water's edge, flying kites, bumping around in go-karts on sandy tracks, some people are even playing golf.

"Look, there's Mario," Silvia yells, pointing and waving back at a man dressed in black.

"He hasn't changed at all!" Alan says, remembering him from the viaduct. Even Nika knows some stories about him, from her parents, about when the height of his work was to entertain people at traffic lights with his freestyle footballing, a self-styled football juggler. He is holding a top hat and wearing a black leather waistcoat over an equally dark shirt, despite the midsummer heat. He has scimitar sideburns and a floppy forelock, the like of which Nika has never seen on a human head before.

He is accompanied by two people and a large dog: a woman with a full head of fire-red dreadlocks wearing a post-apocalyptic survival-style pleather combination covered in pockets, loops, dogclips, and buckles, and a tall, thin young man wearing a yellow and red tracksuit. The dog, on the other hand, a German shepherd with graying fur, has its paws in the water and is sniffing the air, ready to welcome them.

As soon as they have landed, and after the hugging and introductions to Zhenia, his companion, and Romano, his son, and the old dog, named Rocko in honor of Elvis, Alan shoots his first question.

"What is this stink?"

Silvia catches sight of the line of vans and minivans parked in the dunes.

"What is it, are they having a village fete on the beach?"

After sniffing each other, Bielka and Rocko point their noses towards the source of the smell so attractive to both animals and humans.

At first sight, Nika can only recognize a few signs hanging on the parked vehicles: pizza slices, Nutella piadine, panini, assorted snacks, ice cream and yoghurt, seafood, supplì and croquettes, mountain cuisine, mixed fried fish in paper cones, and chickpea farinata.

She finds many of these names impossible to link to anything she knows; the aromas are foreign to her too, a mixture of frying, breadcrumbs, salt, glazing, smoked food, and spices coming together to almost instantaneously overcome her olfactory senses. The interpretation of each olfactory stream is made more complicated by the sweet aromas coming from the suntan lotions worn by the bathers, and hundreds of inhalers scattered beneath the beach umbrellas.

"No, actually it's an antique food exhibition," says Mario. "The local governments of Lazio and the City of Rome like to remind us of our traditions, so they finance this kind of activity at least two or three times a year. In the summer they gather near the beaches because there are lots of people, whereas in the winter they head towards the piazzas in the center of town."

In the middle of all these traditional food proposals there are some nutraceutical classics: bowls of nano-enriched wok-to-walk, pills of high digestibility nutraceuticals, smoothies of hydro/aeroponic fruits and vegetables, hyper-protein sweets, hypocaloric meal replacement juices, and a myriad of DIY print and bite snacks at nanomat stations.

Propelled by curiosity, Nika goes over to a food van surrounded by about twenty people. Her clothes are wrong, even though she couldn't say how or why: out-of-fashion colors or over-simple 3D freeware models without details and finishings.

People smile at her, relaxed, and she feels even more embarrassed when she sees their hairstyles: bright colors, aerodynamic fringes, braids knotted in bizarre ways, layered gravity-defying cuts, zigzag shading highlighting parts of the skull, fluo extensions and shaved areas sculpting geometric lines and shapes, works of hairdressing engineering.

Behind the glass of the kiosk, between the displays of the golden grit used to breadcrumb, there are holographic fish menus in the shape of fans stuffed with wedges of potatoes. The temptation is strong, Nika is about to choose something from an automatic taster. On the ordering column display, there is an antiquated touchscreen and various flavor indications: dairy, grilled, aromatic, sweet 'n' sour, hot, spiced, fruity, and fish. Below is a time option for flavor release: slow, normal, immediate, or delayed release; at the bottom, intensity on a scale of one to fifteen. All of it can be composed within one minute.

The young man running the van is busy serving his customers and Nika is worried he might not want to tell her the list of ingredients, so she selects a box marked 'compositional material ingredients'.

A high, squeaky voice asks, "Ciao! Are you allergic to anything? If you have a nano-alimentary intolerances certificate, I can eliminate the specific problem molecule from the composition."

She takes a step backwards and knocks into the young man standing in the queue behind her. She doesn't need to scan the nano-food to know the crunchy covering of the sandwich he is holding makes up more than half the volume of the food.

"We'd better get moving," says Mario, encouraging the group to head towards the dunes. "It's nearly time for everyone to be going back to the capital."

Nika takes her eyes off the food extravagances and returns to the others at a quick trot.

After walking a few hundred meters inland, clouds of mosquitoes rise out of the undergrowth, huge, hungry-looking flying beasts.

"Ah yes, I was forgetting," says Mario, used to the sight. "There are more insects than starlings in Rome now, it's the climate's fault. Be careful, these beasts are voracious." Having said this, he takes a canister from his shoulder bag and, spraying its contents to the left and right, cuts a path through the cloud.

The swarm is just the vanguard of thousands of midges, mosquitoes, and flies that are on them in a moment, a miasma of insects accosting them like a fleet of attack helicopters.

"Stay close, the spray bothers them. There are enough of them to make herds of sheep run and suffocate cows and horses by blocking their throats and nostrils. In Serra Spino we have put up defenses to protect ourselves but here we are at their mercy."

"Ow! I got bitten!" Nika complains, holding her arm.

In the space of two minutes, they have been bitten on their necks, wrists, and ankles, the bites of thousands of Lilliputian mouths that won't vanish till the morning after.

Over the sweetcorn fields, between AXA and Malafede, there are other swarms, but these are agricultural drones, moving at various altitudes, scanning the automated greenhouses for dry zones or infestations to remove. As soon as they find a place where they have to intervene, a red LED lights up on their shells, and their covers open like floral corollas.

Further ahead, the enriched asphalt of Via Cristoforo Colombo glitters, anthracite colored. The seagulls rising from the banks of the Tiber in search of food often mistake the glittering surface for water; this happens so frequently the road's surface has become a twenty-kilometer-long graveyard for dozens of dead birds.

"Poor things. Why don't they just fly away?" Nika asks, saddened by the scene.

"Because they live here like people do," says Zhenia. "After gliding to the ground, they realize their mistake, but not all of them manage to take off again before being run over."

If flying is difficult, walking is discouraged if not actually forbidden, with every pedestrian path that still exists closed and the entrances barred. It is impossible to cross the fields without trespassing on private property. The signs they come across are not welcoming:

NO THROUGH WAY

NO ENTRY

AUTHORIZED PERSONNEL ONLY

NO THROUGH ROAD FOR HAND- OR ANIMAL-

DRAWN VEHICLES

CLOSED TO TRAFFIC

PEDESTRIANS ON THE OPPOSITE SIDE

NO STOPPING OR PARKING

"There are no shortcuts to Serra Spino," says Mario solemnly, as if it was a test to pass before they could reach the yearned-for urban oasis.

When they leave the maquis in Castel Porziano, the Pulldogs carry on until they reach Via di Malafede and cross Via Ostiense, a knot of streets infested with surveillance cameras every three hundred meters. Not far ahead the Tiber flows invitingly, and some of them want to stop to bathe and refresh themselves.

"We can't park in the open," says Romano, using a verb that sounds strange to Nika. Not so much for the accent her parents still have traces of, but because it is a word usually used in reference to motor vehicles.

"Why not?" she asks.

"Because wandering around is forbidden."

Within a few minutes, they ford the river on a rubber dinghy Mario had left hidden on the shore on the way to meet them. Then, before they can even see the outlines, they hear it. Waves of Doppler effect noise break over them from a hundred meters away. The Pulldogs can no longer even hear the sounds of their own steps. They have to lower their eyes to see their feet moving and not lose rhythm. They have to concentrate to complete such a simple movement in an environment so hostile to walking. For the first time in her life, Nika perceives her body as if it is out-of-place matter.

"This way, c'mon Nika," Miriam shouts, seeing she has fallen to the back of the line.

The grass is the color of rust, the bushes are like wood shavings. Pets gone wild feed on the rubbish littering the streets. The only forms of vegetation permitted are lampposts, signal repeaters, road signs, billboards, and the poles for the ubiquitous surveillance cameras on which the drones have made recharge nests.

"Smile, but don't wave," says Alan to the group, finding his old, sharp sarcasm. "The Rome-Fiumicino motorway is watching us."

Nika is scared. The monstrous tires of her nightmares are spinning as fast as they can close by, she can feel her ears being overwhelmed, the smell fills her nose, it irritates her skin. Every now and then, she walks with her hands over her ears to keep the chaos out. But it is no use, the decibels pass right through.

The Rome-Fiumicino motorway is wide enough to support a six-lane flow of cars. Facing all this rubber, plastic, and metal, Nika wants to hide behind the guardrail with Askalu and watch that daily transhumance. Maybe, who knows, once she gets over her terror she could stand rocking in the emergency lane playing with the rushes of air, enjoy the vibrations and the sight of the colorful serpent with thousands of white eyes and red tails. It is like a boa constrictor tangling and strangling its owners, spitting a stinging substance at everyone, hissing from both carriageways, hooting, accelerating, and stopping suddenly.

For a moment, with a little imagination, she pictures herself on the center island facing a scene like a gallery of horrors: criminal overtaking, rude hand gestures in reply, furious sneers, sleepy trances, alcoholic hijinks in the cars, teleconference arguments, furtive inhalations, pre-

and post-work clothes, a procession of the quasi-living astride the back of the serpent. Then she imagines people sleeping in cars driven by road intelligences, with their heads resting on the steering wheels or leaning back against the headrests, lying on the back seats, daydreaming after nightmares....

The distressing howl of an ambulance siren brings her back to the path they are walking along through the fields. Groups of Ceylonese on bicycles wave to them and carry on cycling. After a turn, the country road pierces the Rome-Fiumicino through a narrow underpass.

"Years ago," Zhenia yells, seeing Nika pale at the sight of the monstrous traffic, "a group of anonymous activists from East Rome released an engineered microbe that fed on tar. Within a few hours the San Cesareo junction became sand; the traffic of central Italy was paralyzed for a month."

"A month?" Alan asks.

"Yup, better than Walk Pride. The authorities arrested the people responsible after five days and had to scramble to fix the problem by commissioning the composition of the low-pressure travel tubes. The perimeter is now patroled by drones and officers with nanotech inhibitor devices."

"Why don't they do that here too? At least it would remove these cars from view," Silvia asks.

"I think the public calculation capacity ran out. Private companies don't feel ready to sponsor Italy's roads. After all the collapses and the hydro-geological instability, the risks are too high. Some politician or other will use the argument as his battle cry during the next election campaigns, though," says Zhenia cynically.

Beyond the Rome-Fiumicino motorway Nika can just about see closed-in and guarded settlements, satellite neighborhoods like Vitinia, Dragoncello, Centro Giano, and Giardino di Roma, with their own soya fields, self-sufficient server farms, domestic wind farms and solar terraces, giant 3D printers, and pyramids of bricks up to thirty meters high. These are the raw materials – undifferentiated and who knows where from – waiting to be transformed into any variety of objects.

The air is full of flying drones. Some, hovering in geostationary flight, are monitoring the traffic, like old-style traffic lights; others are delivering parcels, from nutraceuticals for people who don't want to compose their food, to goods of all sizes and dimensions. Others are collecting tolls from driverless cars, flocks of sheep without a shepherd. There are also surveillance drones defending homes, babysitting drones taking children to school and accompanying them on their daily activities, companion drones looking after the elderly better than zealous domestic dogsbodies.

"Do the drones drive themselves, or do they have some kind of pilot?" Askalu asks, surprised by how many there are. In Sangha the drones came from a long way away, there were foreigners guiding them and to him they looked like bats he couldn't trust.

"It depends," Romano says. "Some people can afford a real person, others have to make do with an AI. In a few years, it will be the other way around. Lots of my friends are part-time pilots during the summer holidays."

On the right, in the distance, an asphalt collar stretches to the horizon, just touching the wavy profiles of *Castelli romani* and keeping the buildings of Rome in. Seeing the level of security, Nika imagines the comings and goings of the inhabitants are regulated by an army of drone-cams and virtual jailers monitoring every evasion, twenty-four-seven, even though in reality people do no more than go out for a bit of fun lasting maybe a few hours, at most a few weeks of holiday.

If the cage is big enough, prison is easier to bear, Nika thinks to herself, turning to feel the warm air of the Rome-Fiumicino run through her for the last time.

She touches her forehead. It's burning.

CHAPTER SIXTY-SIX

Autotrophia

The shady footpath is pleasant. Romano's photosensitive glasses aren't necessary, but he seems to be wearing them for other reasons; as a hobby he's an untagger, by using an A/R app he deletes every digital trace left on the net by the passage of the Pulldogs. As he walks, he picks up and puts down invisible objects, kicks low blows left and right at virtual tags, sometimes he stretches and points his phone at something as if pointing a policeman's Taser.

"Serra Spino will continue to exist as long as it remains dark," his father says, guiding them through a thick bank of brambles, "a discovery for the more courageous walkers. We got rid of every single road sign and created islands of vegetation as camouflage and protection."

"Why all these precautions? We are in the outskirts, not on the viaduct…" says Alan.

"You're right," Mario says, "but we don't want to risk having to leave here too. The falcons protect us from advertising drones, and we asked the town council to leave us alone. It surprised us when they agreed, and they treat us like ethnic minorities, hermits, ascetics, and nuns. Gas, electricity, and running water trace the grid of our exclusion from their services, public or private, it doesn't matter. For the authorities, if you have them you are part of civilization, otherwise you are out."

"I wouldn't complain, it's better than being treated like ants, rats, or mosquitoes."

Serra Spino farm is in a clearing at the top of a slope. Nobody had stayed on the viaduct for long after the wandering group left, and the few who remained went back to a quieter place, less exposed to the attentions of speculators and unscrupulous contractors.

Romano takes off his glasses, only now feeling safe from data tracking.

Some structures on the farm, the vegetable garden and the well, date back to the original building, whereas others seem to have been composed very recently: a solar paint–powered oven and three sunflowers.

At a certain point, the windows of the farmhouse begin to shake, even the roof tiles are vibrating. Askalu and Rafabel put their hands over their ears.

"What's this noise?" Silvia asks, terrified.

"Fiumicino airport's third runway. It came into operation two years ago. You'll get used to it, like we have," yells Mario, pointing at the belly of an airplane brushing the tops of the trees before passing over Serra Spino.

"It's not possible," Alan says, shaking his head.

"Despite all this," Mario adds, "with double glazing, sound-absorbent curtains and armored doors, Serra Spino has resisted for years. It's true, we have to raise our voices and the traffic is more invasive every year, but when we have sufficient calculation potential, we'll be able to assemble a geodesic dome and isolate the hill."

Above Serra Spino, the aircraft crossing the sky come from all directions. It feels like they are about to be attacked by an air force. The vibrations shake their bones and hurt their ears.

"It seems like this territory challenges and scorns the survival of its inhabitants," Rafabel says. It sounds like a condemnation.

The entrance is a copy of the Arch of Constantine composed in white and green with the writing *Serra Spino* above the Pulldogs logo. It is a relic, made sacred by the name it bears. Zhenia hurries the group through. "Let's go in, you'll feel better and you can settle in."

"Take that rubbish out of your mouths," yells Zhenia at Romano after poking her head into the boys' room. "You've become so *greedy*. You chew and chew, but not to fill your stomach, you munch that junk to satisfy your *greed*, not your hunger."

"C'mon, Mum! You're always on at me! Snack, snack, snack...."

Romano is holding a *biscoccino* between his teeth, stuff that is forbidden around here, triple layers of sponge cake filled with custard. Windmilling his arms towards Askalu, he nails him with a series of terrible virtual nunchaku blows.

"Hey, Zhenia, do you want to fight with us?" Askalu asks, out of breath. The virtual long knives he's maneuvering through derma adhesives on his wrists don't take his adversary by surprise. Nika and Bielka are watching the scene without much enthusiasm. She's strumming chords on her guitar, and the dog is gnawing a stick and growling at anyone who comes too close.

"No chance. My battles leave marks on the ground not on a database. Stop now though, it's almost time for supper."

Romano dodges and deflects Askalu's attack with agility.

"Supper? What time is it?" Distracted by the question, Romano pays with his virtual death and lets out a desperate yell as Askalu jumps at his throat.

"It's my turn to cook today, we have guests, so we are all going to eat together. Clear?" Shaking her head, Zhenia turns and leaves.

The two obey, continuing to attack each other with sneaky kicks and pinches. Suddenly Zhenia walks back to them and has another shot. "Oh, and don't show up still wearing those dermas!"

Romano snorts and rips off the suckers. Askalu is a little scared.

"Don't worry about it," Romano says to his battle companion. "My mum likes hunger, but she hates appetite. She says appetite is for people who are sated, satisfied and well fed. It's for people who have nothing to ask of life. If lots of people become like that, if they taste instead of eating, if people pick rather than feeding themselves, we will come to a bad end without even realizing it."

"My mother says something like that too," Nika adds. "Without hunger, we lose direction and purpose."

"They aren't talking about hunger for food," Askalu breaks in, "I know hunger, and it's terrible, it makes you unable to think of anything else."

"Yes, they mean people without stimuli end up being bored and start looking for surrogates which give them the impression of leading a better life."

"Surrogates? I think you talk just like they do," Romano says, turning into the corridor.

"Here they keep saying there are no alternatives," says Mario morosely.

The dinner is made up of dried wild peas, mashed runner and broad beans, and steamed carrots. To drink, they have an infusion of toasted roots: dandelion, chicory, and something Nika doesn't recognize. There is an aftertaste of ashes but it's a little sweet, like honey.

"It doesn't matter what *they* say, what matters is what *we* do. Reality has to be transformed before it transforms us in a way we might not like," Alan says.

Rafabel shoots him an irritated look. It's unlike him to propose such a cut-and-dried mutation. She stares at him as if he's planning some scheme he doesn't want to tell the group about.

"The territory models us more than we model the territory..." Ivan says, staring at a finely composed beaker before drinking, "...until there's no longer a territory to be modeled. 'An autotrophic existence can open a path towards a better life, towards desires that are not the same as simple materialism.' Vernadskij, one of the Cosmists of the last century, said this. The direct synthesis of nutrition starting from natural raw elements available in the territory, without intermediation from the food industry, will radically change the future of humanity."

"Heliophoria hasn't arrived in Rome yet," Mario snaps. "And I doubt it ever will. They already think we are ecologists nostalgic for the bucolic life. There's no way of explaining to them we are not crazy, that we aren't aiming at the extinction of man for the benefit of the world."

"We didn't come back here on an electoral campaign," Alan replies drily. "The only transformations that work are those that change behaviors. But you.... I can understand staying, but why have you never come to visit us? It's been eight years!"

Mario nibbles on some sunflower seeds. He chews slowly, prevaricating.

"Laziness...pure and simple idleness. Nicolas had a vision. You were angry. Perhaps each of you had a good reason for leaving. We didn't. Is that explanation enough?"

"Plenty. But I would like to understand if you might move now that there is a network of communities scattered throughout the world."

He looks at Zhenia. "I don't know...perhaps for a few weeks?"

"Fuck it, Mario, I'm not asking you to become a missionary or a volunteer." Alan stands up and starts walking around the room. "Will you at least give us a hand with the ceremony?"

This time he does not hesitate. "Of course."

"How about if I were to ask you to make heliophoria something concrete and tangible?"

"What do you mean?"

"I need all the couriers...."

Romano moves closer to Nika. She's tasting the mash. "So, where is your nation?" he whispers in her ear.

"What do you think we are, Native Americans?" she answers, frowning.

"Listen, I'm done with eating this stuff. D'you want to go for a walk? These guys are all talk...talk...talk...."

Effectively, Nika is getting bored of all the talking too. She makes a gesture to Bielka, and another to Askalu on the other side of the table and they all follow Romano outside.

In the courtyard, there are people inhaling and chatting. Miriam is talking to old friends.

"Before, when I asked you about your nation," Romano says, "it's only because I've heard them saying that you move from one place to another as you feel like it...."

"Oh, that. Well, our nation is made up of the grass, rocks, animals that slither, animals that swim, animals that run, and everything that grows."

"Cool, how big is it?" the boy asks.

"I don't know. Our nation is where me and my relations are, it is where the stories we tell are, and where we dance and sing."

"How many of you are there?"

"How many what?"

He says nothing, then tries again. "How do you pass the time?"

"The same as everybody else. We have hobbies, we plan projects, we celebrate what there is to celebrate, we play, we do sport, we visit our friends...oh, we study, but in the forests and by the rivers. Should we be doing other things?"

"I dunno, that's why I'm asking."

"Well, now you know," she says, imitating him.

"Here in Serra Spino everybody says you live a better way of life. I was wondering why."

"I don't know what a better way of life is. Imagine leaving your home for a week without money, phone, or food. What do you need to survive well? To be free? We walk from one seasonal nest to the next, two or three times a year. Some years more, some years less. Along the way, we meet people we care about. If you don't like a nest, when you are grown up enough you can leave and find another. If you can't find what you are looking for, you can seek out other people who want the same things you do and create a new one, all your own."

Romano says nothing, listening as Nika continues. "I have some questions for you too...."

"Get on! Shoot!"

"What do you study at school?"

"I go to catering college, I want to be a food designer, even if my mother and my father would prefer me to stay in Serra Spino. Anyway, in addition to programming the Nanocad, they make us study Self-improvement, Financial Literacy, History of Entertainment, Economics of Modeling, Public and Private Digital Law, Culinary Neuromarketing, and Principles of Olfactology."

Nika looks at him, puzzled, and asks her second question. "This is more generic, why do people in the city stay there, instead of wandering the world?"

This time Romano shrugs. "I dunno. How should I know? Every now and then they go *on holidays*."

She turns and shoots a sardonic look to share with Askalu, lying in the grass, his legs crossed at the ankles. Even though Askalu only has a few months of experience with ecolution, he senses the irony in her attitude.

From the darkness of his hoodie, Askalu is studying the comings and goings of passenger shuttles, goods vans, luggage trolleys, and airplane maintenance and stocking vehicles. Everything moves on its own, without drivers. Rafabel comes up to him from behind and puts an arm around his shoulders. "They're not like the buses you wanted to drive."

A rumble rips through the evening announcing yet another take-off. They all turn instinctively towards the noise. The bright lights of Fiumicino airport swallow any glow from the sunset.

"I'm going to bed, guys," Rafabel tells the group. "You should do the same."

As soon as Rafabel walks away, Nika shivers and gets goose bumps. She lowers her eyes as she scratches the irritating itchiness on her forearm, and is horrified to see the skin is rough with what look like scaly blisters.

"Mummy! Daddy!" she yells towards the farmhouse. "My molt is starting!"

CHAPTER SIXTY-SEVEN

Lost in Trastevere

"Are you feeling better, little one?"

Nika shakes her head. "I had a nightmare."

Her bones ache, perhaps they are growing, like spring tendrils stretching out to the light.

"The usual?"

She licks her lips, they taste bad. "No, I dreamed that Daddy went to prison."

"To prison? What for?"

"Because someone accused him of olfactory molestation...."

"What d'you mean?"

Nika blinks, she feels like she's looking at the world through a wide-angle lens, then everything goes out of focus with inverted sparkling colors.

"There was a judge taking a long piece of paper covered in writing from a slot in a wall-mounted computer. Then he read out the sentence and condemned Daddy to two years in an olfactory isolation prison cell."

"Ah, but it was just a bad dream. There's no such place."

Nika pushes with her legs and tries to get up on her elbows, she can feel her skin tight and pulling all over her. The grimace on her face is of widespread general suffering.

"I can't stand it anymore, Mummy...."

Her body is full of pins and needles and flaccid. Nika has been in this tub of icy water for two days.

"Where does it hurt?"

"My bones hurt me when I do anything. My arms, my legs, all over."

These hours of forced rest are as long as the eternity of suffocating summer afternoons, not just because of the heat but also the narcotic, frequently nauseating fumes that have kept her in a sedated state to ease the suffering of her molt.

"It's actually your skin hurting you."

"But I can feel my bones aching."

Miriam pokes her head around the door. "Can I come in?"

"Of course, do you want to check her out?"

Nika's grandmother comes closer and checks on how her granddaughter's epidermis is coming away from her body. The water is cloudy, full of strips and slips of skin floating like in a primordial soup. The molt will reach its conclusion within a week, and after two days of very high fever, gastritis, and acute nausea, Nika has finally stopped whimpering and rambling deliriously.

"You are about to bloom. Another two or three changes of water and tomorrow you can already get up and go out. You won't need anything, except to run and let the air blow over your skin. The sun will make you even stronger and you'll suffer less; actually, it'll make you feel better because the heliotrons and all of your anatomy will benefit greatly from it."

"But I don't want to go out!" Nika complains, lowering her eyes to her reflection in the water. "I can't let people see me looking like this," she says, slapping the water, making waves to obscure the image. Then she starts picking up pieces of epidermis and dropping them back in the water where they float like plankton.

"Don't exaggerate. We have all been through this," says Silvia.

"Yes, us Pulldogs with heliotrons, but the others?"

"So you don't want to come with us to Rome?" Miriam asks.

Nika sulks and slips underwater, head and all. Her grandmother lets her act out, she smiles indulgently and looks down at Nika's face, distorted by the refracting water.

"Don't you understand? I'm embarrassed," Nika says after resurfacing and spitting out a jet of water. Part of her wants to run away from there, from this closed, claustrophobic space, boxed-in on all sides by boring

334 • FRANCESCO VERSO

cement walls; another part of her is scared of showing herself and her diversity in public.

"There's nothing to be embarrassed about," says Miriam. "Our bodies are always changing."

"I know bodies change, but not like this. This isn't just early adolescence!"

"OK, I don't want to try to force you." Miriam touches the back of her hand. Her pictograms are shining with a fluorescent luminescence. "Do you know how I got my heliotrons?"

"No, you've never told me."

"Well then, first I want to tell you something. At six months, a baby koala leaves its mother's pouch and goes from feeding on milk to its adult diet of eucalyptus leaves. But the leaves are hard, toxic, and almost devoid of nutrients. Mammals don't have the genes for producing the enzymes needed to extract the nutrients from eucalyptus leaves, but koalas have solved this problem by using microbes capable of getting energy from the fibrous material of the plant. Unfortunately, baby koalas do not have these microbes from birth so the mothers have to grow a colony of microbes in their intestines, and when the moment is right they produce a meal full of intestinal bacteria which the babies eat. In this way, the little ones get their useful bacteria and develop their first intestinal colony with which they can transform any plant into food."

"That's disgusting! Is that how you got nanites? Through vomit?"

"It wasn't vomit, let's say it was saliva. You are the first person to have nanites from birth, we all had to ingest them."

"Whose saliva was it? Was it my father's?"

Miriam hesitates. Silvia sighs; it has become futile to try to hide stuff from her.

"Not quite. I ate the nanites in a plate of *bucatini alla amatriciana*, many years ago. It happened before I gave them to your father after his accident. I did it to see what effect they would have on me, before giving them to him. On the other hand, I took the heliotrons later from Nicolas when I became a solar plexus...." Realizing the possible misunderstandings, Miriam hurries to add, "It wasn't with a kiss, even though in nature these things are normal. Do you have any idea how many fluids adults exchange in pleasure?"

"Yes, I know that too. It's disgusting!"

Miriam laughs. "It was a phial of saliva! You know, really I have to agree with you, even though it is a universal rite. It doesn't just happen with humans, every species transfers a number of hereditary elements down the generations like houses, money, genes, or nanites. This is why I stopped seeing myself as a person and began to consider myself as an organism formed of many parts, a team of memes, and another of nanites. Like in any relationship, I get from them what I give to them. I provide energy and protection, and in exchange they give me nutrients and support my organic structure. They are my colony, and their survival is worth as much as the healthiness of my cells. Do you see? There's no reason to be embarrassed about the presence of nanites inside you."

A spontaneous smile, curling to the right, lights up Nika's worn-out face.

"All right. I'll try. Can I compose some new clothes for myself though?"

In that instant, Askalu and Romano peek around the door, as if they have been waiting there all this time.

"We're ready," they say in unison, showing her a collection of post-apocalyptic print-to-go clothes on the screens of their phones. Nika raises herself on her elbows to see better, and the boys' jaws drop when they see her nudity.

"That one's perfect!" she says, pointing at a colorful tunic covered in pockets and a pair of running boots with buckles up the back. Then she realizes something is wrong. And it isn't her nakedness. "Where's my bracelet?"

Silvia takes her hand delicately. "Rafabel just borrowed it. She has promised to bring it back. She had to go and speak to someone, and she needed it."

"Who would that be?"

"Your grandfather. Pietro."

In the basket next to the bathtub the shreds of Nika's skin are finger deep.

"I'm going to keep this, a memento," says Miriam, taking out the basket.

From the San Pietro in Montorio viewing point, Trastevere glitters; in the background the ancient scenery of the Eternal City arouses such a sense of wonder in the walkers they have been standing here for the last ten minutes: an open-air museum, accessible to anyone, and composed of amphitheaters, domes, and columns. Every stone, building, and monument oozes history, culture, and art with layer upon layer of luminous holographic icons, their voices whispering information in people's ears.

Myths and legends of the past – from antique Rome, to Medieval, Renaissance, Baroque, and modern times – shine out in a rainbow of signs, accompanying the gaze of whoever has Touristic Artificial Intelligence installed in their glasses, along the meanders of historic experiences, religious festivals, and picturesque anecdotes that even after centuries continue to exert their enchanting power.

Reanimated by interactive gazes or selected by A/R visors, the scenes come to life by the roof of a building, the fronton of a temple, or the facade of a church, reconstructing facts believed to be true and even unfounded hypotheses. In between these extremes, there's an infinite variety of conjectures created by News Jockeys specialized in celebrating the wonders of an ancient time passed to a greater glory: the birth of Romulus and Remus from the womb of the vestal virgin Rhea Silvia; the heroism of Muzio Scevola in beating the Etruscan resistance; the five hundred and fourteen executions carried out under the emblematic red cloak and hood of the 'Boja de Roma', another name for executioner Giovanni Battista Bugatti, also known as Mastro Titta; to the artistic and personal torment of Michelangelo and the sporting endeavors of Francesco Totti, crowned as the eighth King of Rome on the day of his moving farewell to football.

"If it bothers you, you should turn it off," Romano says, seeing Askalu in difficulty. The boy's head is moving jerkily, like a cockerel with too many grains to peck.

"I've never seen so many layers all at the same time. In Bamako they use three at the most."

"The oldest ones are on top," Mario adds, "and they move according to real-time demand. People come to Rome to take a trip to the past, but then they want to do it as if they are in the present.... Tourists are weird, they like to enjoy Roman relaxation, but then they complain if things are slow and expect the height of efficiency."

Hundreds of tags per square meter, nuclei of distributed data, icons of archeological sites and art apps interweave and are animated as soon as a gesture is intercepted – a look, a movement, a step, it doesn't take very much to activate a T.A.I. Askalu and Nika are overwhelmed by a procession of dancing holograms, an itinerant theater so seductive they end up being pulled into thousands of competing universes.

About a hand above their heads, there are shared real-time feeds and posts on social networks; just below, within reach, extra information, interviews, and vintage documentaries fluctuate in the form of colored pills whose size depends on their length.

"Be careful," Romano whispers, resting one hand on Nika's back and the other on Askalu's to pull them away from the parapet. "As soon as it feels like it isn't you exploring the city anymore, shut off the interface."

"Why?" Nika asks.

"Because, as my father says, 'When you look into an AI for a long time the AI is also looking back, into you.' You run the risk of ending up as an element of the scenery, instead of being passersby."

The Pulldogs start walking again. The traffic along Via Garibaldi is silent, there's a buzzing that just about moves the thick, frenetic morning air and driverless cars are running here and there. There are vehicles picking up groups from emergency gathering places, taking the drunks from the night before – the usual tourists who spent the night inhaling on the Gianicolo hill – home, safe and sound; there are young office workers discussing the day's program while each listens to the advice of their remote personal trainers; sleepy students rocking in the bends as they repeat by heart the lessons in their headphones; and groups of tourists going in the opposite direction, channeled towards the 'Fontanone' along A/R routes like herds of livestock towards art-filled pastures.

After a few steps, Nika scratches her nose. Her senses have been smothered and sharpened by the suffering of shedding. In regular gusts, her nostrils detect an unusual smell. To begin with, she thinks it's the exotic taste of coffee, even though a pungent note makes her change her mind. She concentrates on the smell and gets her first response: an extravagant mixture of just-gathered straw and pine needles. How could she have been fooled so? Even Bielka seems confused by all the traces.

Nika rubs her nose three times, wrinkling it as if something has attacked her nasal ducts. It itches inside, deep inside, through her nasal cavity right to the frontal sinuses. It is making her feel dizzy.

"This smell? Do you know where it's coming from?"

Romano is the first to lift a hand, he points at the lampposts along the side of the road. "From up there. Trastevere was the first neighborhood to install perfume poles. They emit smartfumes to neutralize nasty smells and stimulate emotions. The air is still polluted, but you can't smell it. In winter, it smells of holidays. In summer, they prefer to spread sparkling aromas, like high-altitude mountain air."

"Never seen them before. What emotions do they stimulate?" Nika asks curiously.

"It depends on the moment and the area. Some of Dad's friends say there's even an Olfactory Regulation Plan and every municipality is free to decide what to do to its residents' noses. They can breathe productivity in the morning, and sleepiness in the evening, and who knows what else."

"Strange...."

"Nasty, you mean. My father gets angry every time he hears people talking about these poles. Have you noticed the mask he wears? It's different to ours, it is an *anolfactory* glove, it nullifies the effects of all aroma molecules."

"Doesn't he like perfumes?"

"He likes perfumes, not smartfumes. At least, not the city council smartfumes. Once he told me that in Tor Tre Teste they cleared a building of squatters using the stink of decomposing meat. They all ran away within half an hour. Get the point?"

Nika widens her eyes and grabs her mum, who is walking in front of her, by the arm.

"Mummy, did you know about these perfume poles?"

Silvia nods curtly, a meager agreement that is not satisfying. All she says is, "Try breathing through your mouth," and hands Nika an anti-smog mask.

The Porta Settimiana checkpoint – on the intersection of Via Garibaldi and the tight curve of Via Goffredo Mameli – is protected by surveillance drones. The Pulldogs proceed slowly in single file.

"Mario, are they going to make trouble for us?" Alan asks worriedly.

A flock of drone-cams is moving delicately to surround them, silent data collectors for the council and who knows how many other companies interested in profiling the passersby.

"Just don't make any sudden movements. In the center of town we are all monitored, always."

More and more strange things are being added to what Nika has already come across: around them lots of people seem to be literally *wearing* food, some people are sucking at straws stuck in the collars of their coats, others are nibbling snacks they pull out of their sleeves like conjurers; some people are licking their fingers covered in jelly straight from a flavor glove, not to mention the drinks swilling around in liter baby bottles, edible paper wrap cones and the famous *gelgelatos*, all of it being eaten while they are busy doing something else.

As if this were not enough, the clothes they are wearing are carnivalesque; it's like a fashion show of suits and mix-and-match outfits appertaining to the hundreds of urban tribes: aerodynamic jackets and scarves, weather-sensitive boots and bags, clothing for the paramilitary, cardigans with solar paint, *rechargecloth* sweatshirts, revivalist styles from past eras – Baroque, Elizabethan, Moorish – multipurpose backpacks and storage belts, as if it's they who are all really the nomads, unaware of the day, month, or year they might return home.

"If we don't want to be tracked, we use a *homeopathic* diversion," Mario says, pulling a handful of something out of his pocket.

Nika stretches her neck to see what is about to happen.

"Mirror, mirror on the wall..." he recites the magic words as he opens his hand, freeing tiny spheres similar to pollen, "...who's the cleverest of them all?"

When they get closer to the drones, the spores act like fireworks that make sparks but no sound. Bielka growls, suspicious and immobile.

"A swarm of optronic fleas will be enough to protect our privacy. The vibrations capture the attention of the drone and a microscopic mirror on the apex of each flea sends the scan into a loop, emitting a high-frequency disruption signal. The drone looks into the mirror and sees another drone...ad infinitum."

A little further on, all traffic is being diverted and channeled into Via Goffredo Mameli, including pedestrians, segways, and cars.

A six-meter-high holographic Cicerone shows the entertainment and attractions on offer in summer Rome: bareback horse riding courses on Villa Borghese's riding circuit, Urban Golf lessons in the Coppedè neighborhood, nautical licenses in the port of Ostia, free dance exhibitions in the Music Auditorium, motocross, cyclecross, and go-kart races at the Vallelunga circuit, not forgetting paragliding on Mount Mario, skyscraper climbing in the Eur Roma 2 district, parachuting from the Salaria airport, bungee jumping over the Aniene River and wind-surfing along the Tiber, hiking in Villa Pamphili, birdwatching at dusk from the Appia Antica aqueduct, and adventure parks for the big and small at the *Fungo* in the Eur district.

"But they are all activities imitating nomadism," gasps Nika, stunned. They all step down from the pedestrian sidewalk in Via Luciano Manara like a river drawn by the incline of the Gianicolo hill.

"Don't worry about it," Romano reassures her, "it's all fake. The paths are safe, the forests are mapped and there are no pirates in the Tyrrhenian Sea."

"Looks to me like they are suffering from videophilia," says Miriam, looking at how everyone is wearing A/R visors or lenses. "They'll never be nomads, their interest has been deviated *from watching flocks to watching screens*," she says ironically.

Here and there, high-rise kitchen gardens peek out from car parks, roofs, and balconies, examples of urban *agritecture*. The sun hits every surface brushed with convert-energy paint, and in the micro-greenhouses they can just make out the ghost of greenery: chlorophyll capsules, gardening bricks, fruit and vegetable pills all maintaining the illusion of cultivated land, of slow growth: no shade of industrial gray has ever guaranteed comparable satisfaction.

Beneath virtual Rome there's another, less glittery and longer-lasting Rome: cracked roads in the asphalt undergrowth of Piazza San Cosimato, ancient, crumbling restaurants like Corsetti without even a menu pole to pull customers in by their noses, derelict churches that have been closed for decades, like San Callisto and others – as is Santa Maria in Trastevere – taken over by freeware distributors, stalls selling pre-composed Italian souvenirs, and filled with noisy tourists. There are so many *nanoshops*, managed by assemblers who scrape together material from who knows where; nothing like the Composition Salons which sell packets of certified nanites, atomically stable, and, more importantly, without a use-by date. The duration of material is expensive and people who can't afford it have to make do with what is subject to rapid organic deterioration.

"Hold hands now! The lanes are going to get even narrower," Alan yells back from the front of the group.

"From Sant'Egidio onwards there are going to be even more people," Silvia hurries to add.

The crowd appears to Nika as a compact mass of undecipherable odors, incomprehensible languages, loud music, and pushing and shoving from all around. From both above and below, the shop windows along Via della Paglia house *promolograms* twirling like impossible-to-ignore cartoons. Sometimes they stretch over the heads of passersby, brushing them, coming up to them and going through them, whereas others come down to the road to be touched, clicked, perforated, ripped, and hit in points that activate personalized offers on people's visors.

Nika is stunned and confused; tails straight and nose pointing up, Bielka is at her side and Nika rests a hand on her back to steady her nerves. There's so much noise and confusion that she tries to grab Askalu two steps in front of her. Her first lunge misses and at her second attempt, she touches something that is not his hand. It's greasy and sticky.

"Hey," says a woman, licking her fingers. Her T-shirt is similar to the one Askalu is wearing, but she's taller and heavier. "Is everything OK, little one?"

Looking around, Nika can't see any of the Pulldogs. "Yes, sorry," she answers and runs off into the maze of lanes. Barking, Bielka runs after her.

The air is hot and asphyxiating. She is filled with a feeling of fatigue, post-molt nausea, and growing anxiety. Nika opens iMaps: the Pulldogs are immobile dots on her screen a little further ahead along Via della Scala.

She dodges around a group of tourists arguing and shoving each other for the opportunity to get to the discount holograms in front of the augmented shop window of a famous fashion designer.

There's a wall of people blocking her way. Trinket printers, illegal smartfume pushers, street developers of all sorts of objects from prosthetic limbs to glasses and souvenirs.

The icons marking the Pulldogs are moving away. Nika has to hurry. She gets pushed into a luminous circle launched by a salesman in an amusement arcade in Vicolo de' Cinque: the circle tightens around her feet and another around her shoulders. Bielka growls but can't do anything. With the arrival of a third circle – around her waist – Nika wriggles and the hoops vanish. She pushes immediately into the shouting crowd, pulls out her phone and tries to call her mother, but before she can click on the number, two hands push a plate of steaming bucatini in front of her eyes.

"Just made. Would you like to try?" says a waiter dressed like Meo Patacca wearing a floppy beret and a red cummerbund. His goatee is a bit too elaborate; it doesn't fit properly with the Romanesque mask. Nika pushes past him, annoyed, but she turns her head back, attracted to the aroma of the amatriciana sauce. Suddenly she feels dizzy, her knees give way and she drops to the cobblestones. Meo Patacca runs to help her. "Are you all right?"

Nika can't open her eyes, nor her mouth. Her breathing is labored. Bielka isn't there. The man picks her up in his arms and takes her to an inner courtyard, where he sits her on a wicker chair.

"Here, drink this," he says, offering her a glass of water.

Taking little sips makes her feel gradually better. "Thanks, my head, everything was spinning."

"You're not from around here, are you? Rome burns in the summer. You just stay here till you're feeling better."

Nika sees she's at a restaurant similar to Il Romoletto her mother has told her so much about: wooden tables, cart wheels, vintage photographs of 'vanished Rome' on the walls, and a ceiling dripping with strings of tomatoes and dried chili peppers. There are two birdcages in the corners with some silent green parakeets inside; a chained German shepherd whines as it walks backwards and forward in the back doorway; there is also a two-meter-long aquarium with various sizes of fish waiting to become the dish of the day.

Unlike the Romoletto, there are five mechanical arms working in the open-sided kitchen, helped by three food printers. A conveyor belt, like in a Kaiten-zushi Japanese restaurant, sends plates to the customers who then serve themselves. Meo Patacca and a little further on Rugantino, knee-length britches and a kerchief at his neck, are simply agents tempting people in. Then, hidden under a canopy, as if it were a banal ornament, Nika sees there is a tree. It's a splendid bonsai, a juniper growing in a tiny pot, its contorted trunk shaped like a squashed S. The plant is young, but the unhealed cuts left by shears, the anodized aluminum wires wrapped around some of the branches, the cut bark and the concave trunk that has been carved out with clippers, make it look like an old tree that has survived thousands of agonies.

Kenshij told her that originally bonsai were medicinal plants cultivated by Buddhist monks who had learned to grow them in small pots in order to facilitate transporting the active ingredients to treat the sick.

Of the two thick branches coming from the hump of the S, one has been preserved, the other hasn't. The leaves – pruned with maniacal frequency – are green, too green, an indication that someone has been overdoing the water. Raising her eyes, Nika understands why: there's a sign saying:

YOU TOO CAN WATER THE BONSAI
HELP US MAKE IT BEAUTIFUL

She's about to start crying. A watering can with sprinkler is there for whoever wants to participate in this drop-by-drop murder. In an environment with water, fertile terrain and sun, trees grow tall, to the height of their potential, but where the conditions are bad, like here, they lose the impulse to grow, needing all their energy to resist and survive.

Nika is scared she might end up in the same way. She can see it all around her. The smells aren't freely emanated, they are sprayed. Hands aren't shaken, they are grabbed. Glances aren't exchanged, dirty looks are shot out. Steps aren't taken, they are ignored.

In the same way, every plant, before becoming a tree, was a seed waiting for the right conditions for its transformation. Nika, to become a young lady, has to crack her shell and put an end to her dormancy.

This contorted being reminds her of elderly people in the city, their afflicted and desolate existence at the mercy of any tyranny and injustice. The bonsai – an outdoor juniper, no less – shouldn't be inside. Ornamental plants are like caged animals.

"Nika!" She hears a voice behind her and she turns quickly. Silvia and Alan are running to the courtyard entrance, Bielka following behind, tails wagging furiously.

"I couldn't find you," she says, hugging them hard. "I had heatstroke. That man brought me here," she adds, pointing to Meo Patacca, who nods at her parents.

"Everything is all right. I set a proximity alarm on the MyLife365 app," says Alan. "We saw you were missing as soon as you left the perimeter."

"Come, let's go back to the others," Silvia says, taking her by the hand.

After a few steps, they reach Piazza Trilussa. Alan notices a bump on the front of his daughter's tunic. "What've you got under there?"

"Nothing."

He prods the shape and discovers the bonsai. "Where did you get that?"

"I found it."

"Found it? You didn't steal it, did you?"

"No, I freed it."

His look is stern until he smiles smugly, approving the gesture.

Once the Pulldogs have all gathered together by the Ponte Sisto fountain, Alan organizes them into two groups.

"Nika," Silvia says, taking her to one side, "Daddy and I have to meet some friends in Campo de' Fiori, but I've arranged for you to spend some time with your grandmother Anna, she's waiting for you near here." Then she turns to Romano and adds, "Please, can you go with them and take them to the Botanical Gardens?"

"Yes, I'd love to take them," the young man agrees.

"Why can't we come with you?" Nika asks.

"It's a delicate meeting," Alan answers, and she realizes his hand is resting too protectively on his bag.

"Are those the kudzu seeds?"

He bites his lip. "Exactly, but that's all I can tell you."

Nika looks at Miriam worriedly. Her dream from the day before slips into her mind.

CHAPTER SIXTY-EIGHT

The Memory of Smells

The Tiber's Farnesina beach is inviting, with its imported sand dunes and a battery of frondy palm trees transplanted from Civitavecchia's seafront.

Morning inhalations, lunch breaks, and evening sniffing have found their perfect setting, and, judging by the number of people sunbathing, the Romans have rediscovered the pleasure of being able to bathe in the river.

Bielka is excited, so much so Nika has to grab her by the neck because she has jumped onto the parapet. Something is attracting her attention. Leaning out, Nika sees the most bizarre thing she has seen so far: dozens of wild animals running free and chasing each other on the beach. It seems that amongst dog owners there's a fashion like modding, that is, they dye their dogs' fur to make them look more like wild animals. There's a pit bull with highlights making it look like a Siberian tiger; a golden retriever dappled like a jaguar; a spitz colored like a fox; a Great Dane striped like a tiger; a Newfoundland shaved and colored to look like a lion; a chow chow piebald like a panda.

"If they like these wild animals so much, why don't they go and live in Africa?" Askalu asks with a note of indignation.

"I reckon they miss nature," Nika says. "It's such a rarity here, no one gets to experience the real thing, so they make a little fake piece all for themselves."

"But like this they are turning animals into toys," Askalu says.

"There's no nature inside the loop of the ring-road," Romano chips in. "Inside the ring there are only designs…of places, objects, animals, and people. My father says the transformation from wild to design is a measure of identity."

"Now you're the one saying difficult things. How about going for a swim?"

Romano waves his hands to cancel an unwanted promotion from his visor.

"Forget it. I've just been offered a three-for-two entrance to the 'RenaissaRome' beach. They want ten euros just to go in."

"You have to pay to go for a swim?"

Romano shakes his head. "Yes, it's terrible. Anyway, we don't have time, your gran is waiting for us."

Askalu turns reluctantly and follows his friends.

After a few hundred meters, Romano stops to look in through a beauty center window. Two models, a man and a woman wearing only underwear, are displaying various nanite packages with which to execute temporary changes and/or definitive beauty treatments, from hair removal to mani-pedicure, to the regrowth of any kind of hair, to changing eye color and shape, right up to more complex procedures like height modification, facial remodeling, and the reconfiguration of basic metabolism.

"This is where Il Romoletto used to be. Did Silvia tell you?"

"Yes, she told me she grew up here, but she didn't like it very much."

"My father lived in Trastevere too. They met in the San Callisto bar. He started to learn how to juggle and do parkour. She convinced him to sabotage a billboard. Look, I'll show you."

Romano selects a video from his cloud and shows it to the others on his phone.

"In Piazza Navona there was a gigantic installation of a non-modular cell phone, one of those old models that had to be changed every year due to programming obsolescence. The smartphone in the advert was planted like a four-meter-high monolith above a fake cave composed around the Quattro Fiumi fountain, and the slogan said, 'Find the path towards freedom and discover new stories.'"

In the video they see the hooded figures of Silvia and Mario climbing down some scaffolding against one of the buildings in Piazza Navona, from the roof to the top of the phone.

"The next day the slogan had become, 'Lift your eyes from your phones and discover your own path.'"

Behind them, they can hear a humming of voices and overlapping ringtones. Nika raises her eyes to the Gianicolo hill and sees the hefty outline of a modern building looming next to the Botanical Gardens.

"What is that?"

"Medianet, where they film sitcoms and reality shows."

Curiosity piqued, she lets her eyes follow a group of young girls wearing transparent saris, translucent shorts, and colorful chains linking their eyelids to their lips.

"They're fans of the Italo-Ceylonese singer Indurap," says Romano, showing them a hologram of the young man of fifteen intent on levitating in the lotus position over a fountain emitting sounds in time to the spurts of water.

Every time the girls move, they make tinkling sounds.

"What are they doing?"

"Queuing up waiting for momentary luck."

Illuminated routes are traced on the ground, ensuring priorities are respected. As if this weren't enough, a number of guards make sure nobody tries to be clever.

"What do you mean, *luck*? Is it a lottery game?"

"No, they're waiting for a Media Intelligence, a kind of expert filmmaker, to choose the extras and walk-ons of the day and send them onto the set."

There are young men with a retro look, others are wearing futuristic clothes, there are ladies with their faces made-up by online artists, and old people hobbling up waving X-rays and blood reports in an attempt to move up the queue more quickly. Nika picks a flyer up from the ground: entrance, including two nutraceuticals and an inhalation of your choice, costs two hundred and fifty euros.

"Three years ago, my father queued up: he wanted to be the double for Elvis in a TV series about the King of Rock 'n' Roll."

At the entry, three moving walkways take the acting hopefuls, people who have saved from their wages/pensions to participate in their favorite series, to the building for their digital signatures. On the way, they are interviewed by a casting agent with five standard questions: favorite film, favorite actor/actress, what you do best, what scares you most, three facial expressions as preferred.

Hundreds of hopefuls remain outside: people who don't have the two hundred and fifty euros but are trying to raise it by asking for micro-loans through online petitions supporting their cause wait alongside cosplayers imitating the best-loved actors of the moment.

Nika sits down on an old tree stump wedged into a square of earth in the asphalt. Lowering her arms, she touches the sawn-off trunk, brushing the bumpy ridges, the white duramen.

"It's a plane tree, it must be at least three hundred years old."

"They cut them down in the hundreds two years ago after a hemiptera infestation."

"It has lost all traces of chlorophyll and pigment. This tree hasn't been attacked by insects recently."

Fifty meters away, they can see the entrance to the Botanical Gardens. There, technology has protected the trees from the parasites; their canopies rise imperiously, thirty to forty meters high.

"I will never understand why hundred-year-old buildings are considered historical heritage and it's illegal to knock them down, but they can cut down trees that are two or three times older. Poor tree, dead and no one even buried it."

Romano looks surprised. "You bury trees? As if they were people?"

"Of course, would you leave dead bodies around for everyone to see?"

Romano shrugs, almost as if he feels guilty. Then he sees an elderly lady waiting by the side of the gates to the Botanical Gardens. "Ah, there she is, over there," he says, changing the subject.

★　　★　　★

Nika doesn't quite know how to react to Anna's embrace. She stands still, as stiff as a pole, and accepts her grandmother's gesture with a mixture of fear and embarrassment. When she looks at her more closely, she sees a face like the plane tree she has just been pitying, dry and wrinkled, but this woman smells of musk and lavender.

"Veronika. How lovely to meet you. And who are your friends?"

After the introductions, Anna buys them entry tickets while Askalu and Romano take a brochure. The group enter the Botanical Gardens and start following the marked route under the shade of the trees.

"What a strange place. Where I come from people live in tents in the middle of the forest. Here it's the plants that live in greenhouses or are fenced in gardens in the middle of the city. Why did you want to meet us here?" Nika asks, drying the sweat off her forehead.

"I know you love plants, and I thought you would prefer an open-air space," her grandmother answers. "I used to come here with Silvia and Nicolas when they were small. They used to spend a lot of time together after school and they liked to sniff the plants and play hide-and-seek high up there," she says, pointing to the hill, "in the Japanese Garden and the bamboo forest."

For Nika, walking along paths where her mother (and her new father) spent their childhood gives her the same feeling of familiarity you experience when you meet someone you have often heard spoken of, someone like this horde of trees, some young and some over a hundred years old, a small vegetable tribe hidden on the slopes of Gianicolo hill.

Romano gives his brochure back to Anna. "The Botanical Gardens are already a disappointment. It's not like in the brochure."

"Wait before you judge," they hear a deep voice saying. "If you use the app you can live a botanically augmented experience full of stories and interesting information about our plants. We have a plane tree which has been estimated to be between three hundred to five hundred years old, its trunk has a circumference of six hundred fifty centimeters and is forty-three meters high. Next to this is its twin, an imploded plane tree which ended up like this after being hit by lightning in 1990."

The young people download the app and immerse themselves in a hybrid reality using their phones or by positioning their visors over their eyes. Virtual trunks made of infographic film peel away from the trees, icons bloom from the flowers, offering extra information, and guide animals appear in the branches – squirrels, chameleons, and owls – explaining various aspects of the plants.

Once past the Mediterranean Garden and the hedges trimmed into geometric shapes, the group follows a corridor in the building housing the Monumental Glasshouse where the five- to six-meter-high giant cacti instill respect and uneasiness with their surreal, stocky, curvaceous forms, their rough and imposing outlines, their defensive spikes and fleshy flowers; they remind Nika of the Pulldogs for their independence, self-sufficiency, and ability to transform their organisms to survive in any environment, even the most hostile, like Siberia, the desert, or Rome. She takes some photos to share with Shan Jao.

"What class are you in now?"

The question takes Nika by surprise. Her studies are not comparable to Silvia's, or to those of her fathers, or even of Romano or Askalu.

"Well, we don't go to school in a traditional sense. Our classes are paths we follow with the Path Master for four or five hours a day."

"But you do study, don't you? Your parents...I mean...they think your education is important?"

Silvia had warned her that it wouldn't be easy to find a subject they could use to get to know each other.

"Yes, but the system and the teaching are different."

Anna seems just as awkward as Nika feels. She asks no more questions and continues towards the 'Giardino dei Semplici', which is a collection of medicinal plants in brick flowerbeds on a raised area with a pool in the middle.

"Silvia and Nicolas used to spend a lot of time here. He adored these plants and enjoyed creating pharmaceuticals and chemical preparations to take to school as science experiments. Silvia learned a lot from him. I wonder if she still remembers...."

"She does, she told me about a special plant...*liquidambar orientalis*."

"Ah, yes, it's close to the French Glasshouse, let's go back and I'll show you," Anna says.

When they get to a structure full of flowerpots, Anna points to a nearby tree. Nika points her phone at it and an animation tells her about its history, provenance, conformation, characteristics, and the general uses. She notices small cuts in the bark of the tree opposite her, old wounds made by the point of a glass or metal blade.

"With authorization and guided by the gardeners, Nicolas and Silvia used to take sap from the plants and make chewing gum. It was a game, but very educational."

Those cuts, together with the resinous, lightly balsamic smell of stiffening sap on the bark, are windows into the past. Nika can imagine the scene: the cautious gestures of the kids as they approach the plants like they were some kind of aliens and the surprise of discovery. She imagines her parents, still kids, experiencing a contact, a connection that goes beyond time, lengthening it like a rubbery substance, uniting what would otherwise stay separate. Smell is so effective, generating mental associations that solidify memories more densely than experience itself.

"Do you like Rome?" Anna says, interrupting the enchantment of that revealing regression.

"I don't know. It's...so full of cars, buildings, and people."

In the meantime, Romano and Askalu have set off to look for Bielka, who has vanished into the bushes.

"Are you going to die soon?" Nika asks her grandmother without warning. "I read that in cities people live at the most to ninety."

"I am eighty-five."

Her answer pushes Nika to take Anna's hand. She looks up to the underside of the pine canopy. "You see the spaces between one branch and the other?"

Anna nods.

"It's the shyness of the canopies. As soon as the branches sense the presence of another plant – a family one – they stop growing in that direction, so as not to cover one another and steal each other's light. This way they leave each other space to grow."

"What a beautiful thing," says Anna, moved. The image wipes away so many controversies and arguments she had had with her daughter. Nika realizes she might have been sent as an ambassador because Silvia didn't have the strength nor the courage to come. When they lived together, Anna and Silvia cast shade over each other, neither of them knowing how to leave any light for the other.

From the Rock Gardens, Romano and Askalu are waving wildly at them.

Bielka must have ended up in the bamboo thicket there; every now and then they can hear her barks, probably chasing a bird.

While Anna climbs the hill, Nika takes advantage of the moment to take the bonsai out of its pot, dig a hole in the ground and plant it there so it can grow without people watering and pruning it all day long.

Then, from behind a rosebush, they hear a woman's loud voice. "Nika! Askalu! I'm here!" Rafabel is running towards them. Askalu runs down to her too, even Bielka comes out, tails wagging, to greet her.

"Silvia told me I'd find you here," she says, introducing herself to Anna. "I was passing nearby, so I thought I could bring you your bracelet back. Anyway, it's been so long since I came here."

She hands Nika the bracelet and her enthusiasm vanishes.

"Did you meet my grandfather?"

"Yes. It would have been better not to."

"Why?"

"It was a mistake. I recorded a video, though. It could come in handy if he wants to cause us problems in the future."

"Can I see it?"

"No, it's really better if you don't."

"But I've met Anna, why can't I even see what my grandfather looks like."

"It isn't the same thing. Anna is a good person."

"I'm not a child anymore, and he can't be a monster."

"He is, though. I have only just met him and it made me understand a lot of things."

"I want to understand too."

"All right then, little Miss I Want to Understand Too, but don't come to me later and say I didn't warn you." Rafabel takes the marbles out of her hanging earrings, rubbing them to switch them on and project the footage.

CHAPTER SIXTY-NINE

Rome Wasn't Destroyed in a Day

The shot shows the Piazza dei Quiriti in the Prati district. There are eight diffusers mounted in pairs on lampposts releasing fragrances along the perimeter of the circular piazza. Nika can't smell anything, but she senses the presence of colored puffs in the air. In the middle of the piazza, there's a garden with the Fountain of the Caryatids, its central water spout held up by four nude women, gushing water.

"Pietro has patented a model of diffuser and installed them on half the lampposts in Rome," Rafabel explains. "Those lampposts are his pride and joy. He calls them his 'street perfumeries'. Thanks to these lampposts, the city council can guide the moods of the passersby, improve some of their behaviors, like proper attention at road crossings, and discourage others, such as gatherings in certain areas; or they can create pleasant atmospheres, stimulate sociability in the piazzas and well-being in museums, gyms, and shopping centers. It's the perfect 'weapon of mass distraction' – as Silvia called it – discreet and almost invisible."

The cameras in the marbles are focused on the Rendezvous, the venue where Nicolas worked for years. Adrenaline rushes through Rafabel's veins, so much so the video shows the heliotrons on her shoulders shivering; even her hair becomes scarlet red.

The person in frame must be Pietro. Rafabel's muscles are tensed. She's almost trembling.

An overall look shows the Rendezvous holds business brunches, informal meetings around an effuser and an unending number of encounters, from the romantic, to friendly, to family. There's a conference room like a theater for eighty people, two educational rooms for forty people, and another banquet hall for twenty, as well as many eight-person olfactory stations. With the app and a subscription to the Rendezvous, you can go to the Composition Salon on the first floor and enjoy the services: aromatherapy, trimodeling, and the recharge solarium.

The Rendezvous is not a bar and beauty center; it's a whole building of olfactory wonders spread over four floors.

They can see Pietro is in a good mood, he's shaking hands, making jokes with the clients, and patting the backs of the regulars. When he passes Rafabel, he hardly deigns to look at her. He wouldn't even have recognized his son if he had been there, with his sunburned skin, beard, and dreadlocks, even if he bothered to look beyond outer appearances so strange to him.

Rafabel, on the other hand, is studying Pietro carefully. The more she looks at him, the less she likes him. More than anything she doesn't like the smell he's emanating; it's an exquisite aroma with a hint of laurel, reminding her of one of Path Master Kenshij's maxims: "Don't trust anyone who always smells too good."

The ceiling diffusers are pumping an inebriating smartfume over his clients. This week's mixture is an allegorical fragrance of mandarin and basil with a mysterious factor that can be bought exclusively at the aromachology counter. The aromachologist is the new thing, the man who took Nicolas's place, no longer a perfume designer but a specialist aromachology technician.

To begin with, the position of aromachologist was doctor of smells; at the Rendezvous the role became a kind of 'olfactory director', to the point where he sets up perfumed scenery and orchestrates perfume concerts capable of arousing emotions, unleashing memories and *mnemographies*, he conditions the mind using irresistible and indecipherable smartfumes instead of the usual special effects trickery.

There's even music, a languorous funk tango. Three *smelly dancers*, covered in glitter, are dancing sinuously around poles and slithering along the floor, flexible bodies, legs stretched out and spread in sensual poses.

Clouds of incense and streaks of sandalwood wrap around the small pantheon of random divinities chosen from various religions so as not to offend anyone, and in the inner courtyard there are three vases of ornamental plants. The papier mâché elephant Nicolas had told her about is no longer there; in its place there's now a votive altar with a 3D-composed statue of Nicolas, or rather the Nicolas his parents wanted him to be, an imaginary extension of his father's ambition mediated by the protective anxiety of his mother, and an urn, empty but symbolic, where people can leave offerings.

On the side wall, a holograph shows photos, video clips, audio clips, diplomas, and prizes won by Nicolas. Everyone is free to upload something (rendering them eligible for a discount at the checkout), or leave a digital signature and compose a personal memory. The Tomei family owns a web domain that doesn't end in .com or .org, but .family. For Pietro, this must seem like an excellent idea for preserving his legacy: centuries later, future generations of Tomeis, and strangers, will be able to connect to the site and learn about their endeavors. A posthumous, potentially eternal, glory.

It's the moment. Rafabel closes in on Pietro and introduces herself.

"Mr. Tomei? I am Rafabel Cosser, a friend of your son Nicolas. Can I talk to you?"

Pietro is surprised and suspicious. "It's been more than ten years since I last saw him. Don't you think it's a bit late for him to come back?"

"Well, that's what I wanted to talk to you about. Is your wife here?"

He looks dubious. He can tell this is something serious.

"Wait for me at that station over there. I'll go and call her."

Pietro comes back after two minutes, followed by Olga, a woman with sweet but time-faded eyes. Her clothes are excellent quality and rustle expensively as she sits on the sofa.

"Good evening, Mrs. Tomei, I'm a friend of Nicolas and I've come to bring you some news about him. How long is it since you last spoke to him?"

"He called one day many years ago. A few weeks after the birth of his daughter, I think he was in Turkey. Then I didn't hear from him much. Perhaps once or twice, from India, and Egypt.... He said he preferred to stop communication, to avoid suffering on both sides."

Olga is holding her hands in her lap. A posture suggesting the creation of a barrier, but it is also a maternal, sincere position.

"Well, if you want to, you can meet your granddaughter, Veronika. She is eight now, and in Rome with her family. On Thursday we will all be coming together for a ceremony...in remembrance of Nicolas."

Olga puts her hand over her mouth and closes her eyes.

"Oh, I'm so sorry, I thought you knew...that the statue was a way to...."

In the view her earrings permit, they can see the courtyard. Effectively, judging by the dates on the posts and signatures Rafabel hadn't noticed, the statue dates back to at least eight years ago, and therefore couldn't have been erected as a memorial, it was taking advantage of his absence and exploiting it to induce people to leave donations.

"My wife is a bit emotional. We aren't completely unprepared," says Pietro without turning a hair. "Considering the life he chose to lead, it didn't take a genius to realize he couldn't last for long. Just help me understand, what ceremony do you want to hold? It isn't a funeral, is it?"

"We're going to gather at the Acatholic cemetery in Testaccio. We will take a walk through the graves and each of us will remember him with a short speech. Nicolas would want something like this, he wouldn't have wanted an official funeral, too sad, too serious."

"And what would you know about what he would have wanted? You can't do whatever you fancy with our son. We are his family."

"With respect, you *were* his family."

"How dare you!" splutters Pietro, anger rising. "Family isn't a walk in the park, it isn't a thing that starts, has a beginning and an end, nor is it a game you can enter and leave as you like."

"I agree, but as far as I can see, you didn't wait long to compose a votive altar for your son, as if he had been dead for years already."

Pietro lifts a finger and is about to launch a counterattack when Olga stops him.

"Please, what time is this ceremony going to be?"

"At four thirty. Listen, I didn't come here with the intention of fighting. I respect your relationship with Nicolas." Rafabel opens her backpack and pulls something out of it. "I brought you his bandolier as a token. Nicolas used this cartridge belt to compose extraordinary objects that have helped so very many people."

"Thank you, we'll be there," Olga says, taking the belt, which quickly ends up in Pietro's hands.

"The belt is empty," he says after handling and studying it carefully. "There are no phials, nor formulas. This isn't my son's work, it's just a simulacrum. You kept the real one, didn't you? Those formulas are ours; they belong to the Tomei family."

"The formulas belong to everyone. That's how Nicolas did it. Check out what I'm saying, they are available as free downloads on e-Den. What you have composed in the courtyard is a different Nicolas, your version."

"It doesn't matter which version it is. I am his father."

"You are not listening to me, just like you never listened to your son."

"You are the one not listening to me now! We are not talking here about the relationship between a father and his son. I have looked into what you are all doing. It's reckless and dangerous. Does my granddaughter have *nanites* inside her?"

"Your granddaughter is really well. As healthy as can be."

"Perhaps I haven't explained myself properly. Nanotechnology is one thing, as you can see, I successfully use and sell at the Rendezvous, but nanites are another thing. If they can reproduce – if they have passed from my son to my granddaughter – it means the process of natural selection has changed. And if the nanites reproduce, including through imperfect copies, it means they are evolving. Reproducing evolution with artificial replicators is reckless and *very dangerous*."

"Dangerous for you. Because you think it has to be nature as we have always known it that regulates evolution. Now we know how evolution works, we cite nature to maintain the status quo. But what do we really know about nature...it might easily have been using humans until a certain point of evolution and might now prefer to use other vectors, like nanites, to push ahead with another phase."

"What are you talking about?"

"I'm saying the value of uncontaminated, non-manipulated nature is false. Nature is hybrid, and subject to continual mutations. Every organism has to learn to manipulate nature to survive, and every organism, including humans and nanites, obey the laws of nature when they carry out these manipulations."

"That's a load of rubbish."

"Your son wouldn't have agreed."

"Nanites don't exist in nature. They are artificial and have to be managed by people who know what they are doing, not by a group of degenerates like you."

"Degenerates like us? We are the reaction of nature to your mania to consume the world with industrialization, pollution, and commercial exploitation of anything and everything. Climatic changes, the extinction of wild species, the invasion of immune-resistant insects, who caused all of these things? Up until now who has interfered with nature without anybody objecting? Is chemistry natural? Nuclear fission? Do they not have an impact on evolution? But now you are scared of the consequences? Well, it's too late. I'm not even sorry."

"There are billions of nanites. Their out-of-control exponential increase is not manageable. The plagues of rats and locusts are banal catastrophes in comparison."

"In the same way there have always been bacteria in our bodies. We have learned to live together."

"You are putting an end to the memory of thousands of years of history enclosed in biological DNA."

"On the contrary, we are giving it an extra tool for surviving the disaster you have caused and don't care about. You are so blind and hypocritical, you can't even see you are defending the purity of a cadaver, with your cities, your factories, your motorways, shopping centers, airports, theme parks, zoos, and banks. What DNA are you talking about?"

He rolls his eyes, intolerant of clichés and New Age propaganda.

"You are worse than that other madwoman who came back to ruin Nicolas. I hoped he had got free of her, but she got her claws into him again and dragged him onto your viaduct for mad people."

"That other madwoman is the mother of your granddaughter."

He blinks. He has only a moment of hesitation before sneering.

Rafabel can contain herself no longer. She stands up and moves to leave.

"What you have built up can never exceed the value of what you lost," she concludes to Pitero's face. She turns to Olga and asks, "Can you tell me where the bathroom is, please? Then I can find my own way out. In any case, I hope to see you on Thursday…for Nicolas."

Rafabel stops the film.

"Then what happened?"

"I really can't tell you that."

"Are you starting on that too? Is it anything to do with Alan's business?"

"Yes, all I can tell you is that though Rome wasn't built in a day, it can be transformed in a few weeks."

CHAPTER SEVENTY

Earth Invaders

"Once, on the island of Büyükada, Nicolas told a Turkish vlogger who asked him if the Pulldogs were a kind of sect brainwashing people to convince them to use nanites, leave the cities and dedicate their lives to economically irrelevant activities, 'Perhaps we started off as a kind of sect, and perhaps we are a cult now, a movement that might one day become a culture, who knows…in any case, we are a do-it-yourself culture, in the sense that the Pulldogs wash their brains themselves.' The vlogger didn't understand to begin with, so he, Nicolas, added, 'Not only do we wash our brains ourselves, but we look after the land we travel across, we open walkersways, shared greenhouses anyone can stay in, off-grid sunflowers. We share compositional formulas with whoever wants them.' This radical approach, self-supporting and participatory, is what I like to remember about him," Dikran ends, slowing his pace until he's back with the procession of people walking along the footpath.

Outside the Acatholic cemetery in Testaccio, thousands of people are waiting to enter in turns, in groups of forty. The usual maximum number of twenty has been waived. The florist's hut that sells olfactory palliatives and floral *mnemodours* has been overwhelmed and the manager from Sri Lanka, who usually passed his time playing online with his relatives, is having trouble serving everyone.

In the flowerbeds to the left of the entrance, a statue has just taken shape. It's the work of the Master, his way of paying homage to Nicolas. He sent three fabtotums that during the night composed a 1:1.000,000,000 scale copy of a nanite made, literally, out of hundreds and thousands of grains of whole rice. The nanite is brown, six meters high, and everybody is free to take a handful until there's nothing left, including the edible fabtotums and reservoirs made of starch, vitamins, and minerals.

Walkers have come from all over the world, people who have thought up various activities to say #thanksnico as part of today's celebrations: themed smartfume inhalations, online discussions on ergonet forums, olfactory concerts, flash mobs in front of compository salons and, of course, the switching on of effigies of Nicolas as 1kW man to hold until they combust. Thousands of people are doing the same along the walkersways across the world, people who weren't able to come to Rome.

"Today at six p.m. at the Palladium," says Ivan, reading from his phone, "the Nose Noise Band will play an olfactory concert and then come outside to spread their notes and fragrances through the streets of Garbatella. At eight p.m. Society Robot have organized a performance, *Walkingdom*, an installation of more than three hundred walking drones, an interface between technology, science, and art in Villa Pamphili. If you have other ideas, share them with the hashtag on Global Walker."

The lawn in front of the Pyramid of Cestius is brilliant green, even during the summer.

With one hand, Nika strokes Bielka, and with the other, she holds Silvia's tightly. She has never seen a cemetery before, at least not one with hundreds of graves, one on top of the other, decorated with angels, tombstones with inscriptions in unknown languages, stone busts, commemorative flowers, wooden crosses of all shapes and sizes, and little temples supported by columns like miniature mausoleums.

"Like Gregory Corso, Beat Generation poet, wanted to be buried next to Percy Shelley," says Miriam, pausing in front of the grave of the English poet, "I think Nicolas won't disdain resting next to John Keats, who studied to become a chemist but is remembered as a poet, and next to Bruno Pontecorvo, the elementary particle physicist."

Beneath the pines shading the poet's tomb, there's a sign:

IT IS SEVERELY FORBIDDEN TO SCATTER FUNERAL ASHES IN THE CEMETERY WITHOUT WRITTEN AUTHORISATION. ANYONE FOUND CONTRAVENING THIS ORDINANCE WILL BE REPORTED TO THE AUTHORITIES.

Miriam nods at Rafabel, who hurries to the head of the group.

"When we are born, we don't know what love is. Need and desire, yes, but we aren't provided with an understanding of love at the beginning. Love, like other complex emotions, is learned through living. Luckily, many people, after many errors and a little experience, learn to understand how it works. Love is a virtue we discover when we are young and learn to develop during our lives. If nobody teaches you about it in time, if during your childhood or adolescence you don't find good teachers and don't make the right choices, you risk not being able to love. Nobody. Ever. For any reason. I was running this risk until I met Nicolas." Rafabel pulls a phial out of her pocket, inside there is a seed. "Your body is resting in Sangha, in Mali, and is transforming into a baobab, your ideals have been scattered everywhere you left a perfume, a formula, a composition, but this land – Rome, where you were born and raised – never knew the other Nicolas Tomei, the tireless walker you became, the one we all remember."

She opens the phial and drops the seed into a small hole Dikran hurries to cover. "Your tree isn't the baobab but the black hellebore, because you were the best antidote to the craziness of the world."

Some people are crying, others are holding hands, others have their arms around each other. The Pulldogs gather in silence, they exchange a few words, share a quick glance, a sideways look. The procession moves forward, slowly climbing to the higher part of the cemetery. From outside they can hear a buzzing of people, the ladies and custodians who take care of security are not finding it easy to stop them. No one was ready for this number of visitors.

In front of the tomb of Percy Shelley, Alan surprises them by taking his turn to speak.

"What will the world be like without him? Where will his message take us? Where will our next inspiration capable of transforming humanity spring from? It's our turn now! This is what he would have wanted: we are the people who build paths where before there was nothing, we are the people who share formulas for improving life, who create the art of the future, who compose the notes of our melodies, who exchange gifts without personal profit, who go everywhere without leaving a trace. Now that Nico has gone, it's our turn. We who are and remain Pulldogs. We will see each other around the world, because all the paths that divide us, bring us...."

Suddenly they can hear excited shouting coming from below. "Let me pass! I have to get through!" Pietro Tomei, red in the face and escorted by two Nans agents, is pushing his way through the crowd. Behind him, they can just about make out the shape of Olga.

"What's this farce, eh? A funeral without a body? What ridiculous freakshow have you invented?"

"Nicolas hated funerals," Rafabel says. "If he could have chosen, he would have wanted his death to be celebrated joyfully, a walk like this one, in a park full of fragrances, with people playing music and dancing, like we will later. No coffin, no Mass, no praying, no official speeches. You never knew your son."

In the meantime, other people have begun to gather at the entrance gates. Some are pushing to get in, others are competing to show their gratitude, waving effigies, singing Nicolas's name in unison.

Nika keeps a little to one side, half hidden behind her mother.

"When you came to the Rendezvous, you brought me an empty belt. Where are my son's formulas?"

"I will tell you again. The formulas are available for free on e-Den, sue them."

Olga is studying the Pulldogs, one by one, looking for her granddaughter's face amongst the group; then she catches a glimpse of her. "You are Veronika, right?"

"Yes," Nika answers uncertainly.

Pietro turns and just has time to notice the bracelet on the young girl's arm before she pulls it behind her back.

"Show me that," he says, trying to grab her hand, but failing because Silvia comes between them.

"Don't you dare touch her," she replies threateningly. She really wants to land him the punch he so deserves, but the gesture stays on the tips of her fingers because Nika is looking at her with a scared look on her face. She has never seen her mother so tense, on the edge of exploding.

"So I guessed right. The formulas are inside that. The bracelet is mine. It is compensation for what Nicolas gave away for free after being trained at the Rendezvous. He had a contract to respect, he wasn't free to do whatever he liked."

Nobody moves. The noise from outside is getting louder.

"You may proceed," Pietro says, making way for the Nans agents so they can confiscate the bracelet.

Just then, the entrance gates fly open and the enthusiastic crowd pours into the cemetery. The more excited among them run like mad to get a place, the others follow suit, without even knowing where Nicolas has been buried. In consequence, thousands of people flood the footpaths, trample the flowerbeds, and walk over the graves.

Alan grabs Nika by the arm and takes advantage of the confusion to lose themselves in the crowd.

"Stop them! They have my formulas!" Pietro screams, jumping off the path before he gets knocked over.

The Nans agents try to chase them, but they are slowed down by the not totally unintentional pushing and shoving of the crowd, and after a little way they are forced to give up.

There's nothing left of the rice nanite at the cemetery gates except a carpet of grains on the ground.

Nika slips a hand into her pocket and pulls out a handful of rice.

"Do you want some?" she asks Askalu and Romano, who are running alongside her.

"Yes, energy," Askalu answers ironically, and takes some to share with Romano.

The Pulldogs are running at the speed of an urban gallop, weaving between driverless cars, incredulous pedestrians, and architectural barriers: from steps to railings, bollards to uneven sidewalks and holes in the roads.

"Why are we running? We've lost them!" Nika asks her father.

"We have to get away as quickly as possible. Not because of the Nans either."

"Is it that secret thing?"

"Yes, we've scattered thousands of kudzu seeds all through Rome. A carefully aimed and coordinated distribution. A wonderful action of disturbance, something to break the urban tissue, to fracture it, slow it down, and maybe even destroy it a little. Nature reclaims its space and we are its seeds, pollen produces new life, it slips through the cracks, grows from drains, flowers from the sidewalks, invades the piazzas, and pulls down shopping centers. Nature, even modified nature like ours, is a necessary thing, even when it hurts."

Rafabel draws up alongside Nika.

"I did the same thing at the Rendezvous when I went to the bathroom. I pushed a handful of kudzu seeds in the drains. In a few days' time, the roots will spread through the sewers and come up in the whole venue. Pietro deserves it."

Rafabel lifts a hand and Silvia gives her five, like two kids who have just had their revenge on a bully.

Nika has studied the incredible rhizomatic growth of kudzu. She looks in front of her, the cobbles along Viale Aventino with its ancient buildings and the ruins of the Circus Maximus and the Colosseum in the background, and imagines what will happen when the roots come up through the sewers and drains. If nobody – as is probable – digs a trench limiting the expansion of the kudzu with plastic sheeting and other methods of protection, within a month Rome will be unrecognizable. The effect of the green invasion will be stupefying. The Aventino and Porta Metronia areas, which her grandmother has told her about, will become wild as soon as the roots break the blocks of asphalt as if they were pieces of cardboard. The kudzu will bring disorder and green by surprise where no one has had hopes of seeing it for over a hundred years. Nika remembers seeing online models of Rome made by Giambattista Piranesi in which luxurious nature embraced the ruins of Imperial Rome; abandoned temples had grassy floors, crumbling columns were covered with flowers, and marble statues were wrapped in ivy. The scene will be of similar proportions, if not better. Car parks will become thriving non-places and cars will be part of the ecosystem; no longer being able to circulate, they will become biomechanical sculptures with the bodywork and the interiors ready to house nature. Colonies of insects will penetrate the upholstery, birds will make nests on the dashboards, bright green and mud brown will take the place of all paint jobs, from metallic gray to fluorescent blue. Their lights will be eyes; the side mirrors will be ears; front grilles, mouths. Cracked windows will be woven with cobwebs, and compact layers of decomposed and crumbled leaves will accumulate on the bonnets making comfortable mattresses where anyone can rest. The asphalt will crumble into sticky black granules as if reduced to coal. Spikes of grass will invade the road surface, like hair on wrinkled skin. The destructive voracity of the plants will provoke prodigious ecolution.

As they reach the area around the Colosseum, Silvia slows the pace so they can rest on the Caelian Hill. "Well, what do you say?"

The view of the most famous amphitheater is amazing, it's just a pity it is surrounded by tourists camping in the shade of the permanent tent structures: a pity about the remote excursions executed through flocks of drones which detract from contemplation of the monument; it's a pity about the insolent and insistent T.A.I. advertising every kind of augmented experience.

"How to ruin the ruins..." Nika says.

"It's funny," says Silvia. "People living here don't realize it because they've got used to it. Like fish don't know they are swimming in water. But if you look at a city carefully – even a beautiful one like Rome – you come to realize it's the most advanced form of desert."

Then it happens: Nika breathes in deeply and everything on her sensory horizon ends up in her nose. She sniffs the image of the Forum, the clothing of the hot tourists, the deafening noises of the road, the sweet flavors of the gelgelatos, the sinuous movements of the birds, the tightened muscles of the rickshaw runners, the fixed smiles of the group selfies.

In exchange, she receives total synesthesia, not a photo, not a video, but something like a *rhinograph*: a cloud of impressions like ink swirling in a glass of water, it remains suspended, without vanishing, a scenographic smell-still, a wide-angle emotional print and at the same time detailed down to the last effluence of organic, inorganic, and hybrid matter.

She, having grown up amidst the autumnal forests, the flower perfumes of the spring, the glacial silences of the winter, running over meadows and swimming in rivers in the summer, as well as possessing a natural talent for analyzing smells – inherited from Nicolas – has developed another sense, a sum of the others and a powerful tool for deciphering and comparing dozens and dozens of types of olfactory molecules.

The molt has left her a gift, or perhaps it is another nightmare, but not a nocturnal one: with her sense of smell augmented by the nanites, she can capture the essence of anything and everything. She is now experiencing Nicolas's olfactory maps, merely an image on the phone screen until now, for herself, with her own eyes, ears, and nose.

CHAPTER SEVENTY-ONE

You Will Know Me by My Tracks

"Every family has its legends. Mine are Alan and Nicolas, my two fathers," says Nika, digging two holes in the ground. She puts seeds in them, four grains of maize in the first, and four of hellebore in the other. "Alan is rhizomance, Nicolas is heliophoria. The two things put together make ecolution. If their history can convince people to walk more, that would already be an achievement."

"They will be wonderful stories," Tasia says, drinking from her water bottle.

During the afternoon a strong rain shower dampened the ground, but then the sky cleared.

"In two weeks," Pino adds, "the little plants will be setting their first shoots and peeking out of the ground."

Romano helps her cover the holes with fresh earth. "Anyone who comes by here will see it in full flower. It's a good way to celebrate them. Where are we going now?"

Pino and Tasia wait for Nika to answer. She scrutinizes the undergrowth, looking for snapped twigs, piles of wood, and leaves that have been trodden underfoot. It's an instinctive gesture because she knows how pointless only trusting to sight for finding the path is. Nature likes hiding tracks quickly. Even Bielka is sitting and waiting, her tails sweeping the ground.

"Rhinography tells me north," she says, looking at her bracelet for safety. She still doesn't completely trust her sixth sense and needs the confirmation glancing at the phone gives her.

Since she was small, Miriam has taught her that from a footprint in the grass, the mud, or the snow, you can trace who it belonged to as if they were still here. As organic matter begins to contain nanites, it's possible to recreate something of who went by here from their tracks. However, no other sense could have described Nicolas's reality than the sense of smell. With hearing you can listen to the tinkling of water in a river, the lapping of the sea, the roar of a waterfall, the dripping of dew, and the blowing of the wind, and with touch and sight you can follow the contours of a rock or make out footfalls on leaves, but those tracks might belong to anything, whereas with smell....

As the sun goes down and fills the valley with dappled light, the birds start singing, calling out from where they are hidden in the branches: twittering and whistling, solo songs but sometimes interwoven. Could these be the same sounds her father mentioned in his diary? Nika is comforted.

The guano smells like that on the map and the singing sounds familiar, providing other clues to his passage. In particular Nika has seen a little bird, full chested and proud, singing with ardor in the leafy branches of a tree. She imagines it as a messenger, sent by Nicolas to watch over her, or it is him, incognito, encouraging her to carry on and never stop.

"I wonder what the view is like up there?" she asks her companions. The heliotrons of her father are calling to her: an intrinsic correlation whose strength extends over time as well as space.

"Do you want to compose the nest already?" Romano says. Even though he's sixteen, this is the first trip he has made alone, without his parents. He wants to be up to standard, test his limits, discover his fears.

On the crest of the Apennines, Nika can make out figures made by wind erosion, profiles reminiscent of letters, animals, human faces; she can't help but see signs, and messages, and therefore give them an almost prophetic interpretation.

"Let's get to there," she says. "Bomarzo shouldn't be too much further on."

Tasia and Pino approve her choice.

Even though it's invisible on the ground, the path recorded on her bracelet is scattered with cairns, garlands of dry flowers, and cuts in tree trunks. Nika follows it as if she's sniffing out the tracks of an escaped animal; as if those clues form a complex trail of smells capable of reaching her from hundreds of kilometers away, as if that unique molecular combination could last decades, the same as itself and he who composed it: this is the imposing call she cannot ignore.

She finds herself following the same route he made. There's no melancholy in her steps, no nostalgia for his absence. Every hundred meters a vision, an episode bringing up memories that aren't hers: infusions laid down with care, rocks aligned to indicate a direction, pallets in tree trunks used as shelter.

"Here will be fine," she says, dropping her backpack.

The others open their fabtotums and begin to compose the tents, while Nika, with Bielka by her side, explores the area around the beeches looking for produce. The rhinograph flashes and her bracelet's display shows there's a very strong correlation. She breathes in, collecting clues.

A chiseled stone is sticking out of the grass like an antique flower. It looks like a gravestone without writing, able to suggest memories and fragments of past stories. The man who knew those stories can no longer tell them. Nika tries to guess at Nicolas's day: he must have been looking for the direction to take too. He was walking with the Pulldogs, but knew that sooner or later he was going to break away from them. Nika scrolls through the bracelet in search of other indications. The days on the diary slide past.

7/1/2032. Photo of redcurrants, infusion of chestnuts from Mount Cimini, lavender essence, hazelnut-based supper.

8/1/2032. Text from the diary of Henry David Thoreau: "There is a higher law affecting our relation to pines as well as to men. A pine cut down, a dead pine, is no more a pine than a dead human carcass is a man. Can he who has discovered only some of the values of whalebone and whale oil be said to have discovered the true use of the whale? Can he who slays the elephant for his ivory be said to have 'seen the elephant'? These are petty and accidental uses; just as if a stronger race were to kill us in order to make buttons and flageolets of our bones; for everything may serve a lower as well as a higher use."

9/1/2032. Carved on limestone: "I continue to walk towards the sun in order never to see it set."

Nika lifts her eyes, the last of the sun's rays hit the tips of the tallest trees, turning them into flaming arrows reaching into the sky. With wet eyes she hurries to get back to her friends and the nest.

Alan and Silvia call every two days. Mario and Zhenia even more often.

"Where are you?"

"Have you crossed paths with anyone?"

"Your GPS is off. It's dangerous."

"Leave them alone!"

"They're doing well!" they can hear Miriam saying every now and then in the background.

"Don't worry," Nika says to them every time they speak. "The walkersway is safe, there are shared greenhouses and sunflowers every thirty kilometers. The Francigena Way passes nearby here, the path is a UNESCO World Heritage Site. There are groups of tourists all over the place, cyclists and hikers. Tasia and Pino know the way. How is Askalu?"

"Well, he's taken the place of his brother," Silvia says. "He says the rickshaws in his hands will make sparks. He will call you soon, you should see how busy he is."

"And Rome? What has happened with the kudzu?"

"Mario said they are having trouble getting rid of it from lots of neighborhoods," Alan chips in. "The trimmers are working to a continuous rhythm, but they have to go back again and again every time new leaves appear to stop them providing reserves of nutrition for the roots."

Behind them, on the screen, they can see the sea. Alan wants to leave for South America and take his concerts where music has always been an accompaniment for rites and sociability. Actually, it might be because he has been identified by numerous surveillance drones and is therefore officially wanted for 'attack on the public health'. He only came to know about this because Ivan sent him a link to the TV news from the previous day. On TV he was accused of damage, aggravated by article

425, citing involvement of public buildings, monuments, homes, water and electrical systems, and shops.

"In other places they use herbicides, but that needs special authorization because the chemical products are risky and in trying to eliminate the kudzu they might end up hitting all the plants in Rome. The hardmen from the inner city want it eradicated: they call it 'accelerated extirpation', they cut the branches and then rip out the roots using diggers and pickaxes. In the most difficult points, where the roots are deeper and there are no archeological remains or excavations, they have scarified the ground down to three meters."

"That's a lot of work...."

"In the outskirts things are different, though. They are not fighting the kudzu there. They are using it for grazing: they have goats, who go at it with conviction because they have been raised with other, more appetizing species. And that's all...for now."

"Tomorrow we are going to start climbing the Apennine crest. Right now we are in Bomarzo."

"It was our fourth stop. An incredible place, if the monsters don't scare you."

"Monsters! They are only statues," she says back. "Oh, yes, I have a present for you. Seeing as how the shell of your guitortoise has been ruined, I was thinking you might need a new instrument." Nika takes an object and starts singing 'Il Mio Canto Libero' by Lucio Battisti.

"Does it only have one string?"

"Yes, it works like a harp. To tune it you have to regulate the position of each peg. You see here, the string goes around the grooved bearings, they are mechanical tuners. The tuning pegs are attached at these points," Nika says, showing him the various parts. "I printed it in twenty minutes, but it took three weeks on the road to find all the materials. I'll send you the formula, then you can compose one when you want to."

"Thank you, it's a stupendous present. I'll be needing it. We love you, you know that. Please, show yourself on Global Walker every now and then."

Nika concentrates and, under her shirt, the heliotrons light up. Somewhere along the Francigena Way the shiny corolla of a sunflower turns towards her, switches on and transmits.

She moves to put her headphones on.

"You've got something stuck near your ear," Silvia says.

Nika brushes her fingers behind the lobe and feels a flap of skin left over from the molt. She pulls it off and rubs it between her fingers until it crumbles. Bielka comes closer, sniffs it, then goes away again.

"I love you too," says Nika in goodbye, then she closes the connection and looks at her bracelet.

Only five thousand five hundred more steps till they reach water.

376 • FRANCESCO VERSO

ACKNOWLEDGMENTS

The idea for this novel, and *The Roamers*, was conceived on a train from Rome to Milan. It arose from a conversation with Gianni Loconti of La Sapienza University, a great enthusiast of all things related to the future. His heartfelt call for, "Nanobots for all!" became the cry that gave birth to this entire story. However, without the invaluable support of Stefano Scalich, Giammarco Raponi, Francesco Mantovani, Francesco Grasso and Clelia Farris during the creation and revision stages of these novels, *The Roamers* and *No/Mad/Land* would have remained motionless.

ABOUT THE TRANSLATOR

Sally McCorry was born in London in 1970. Brought up on a diet of *Doctor Who* and *The Hitchhiker's Guide* to the *Galaxy*, she moved to a medieval town in Liguria, Italy, in 1990 and graduated to reading *Dylan Dog*. A fan of science fiction and fantasy from an early age, she finds it deliciously ironic to translate it while sitting by a wood-burning stove.

ABOUT THE AUTHOR

Francesco Verso is a multiple award-winning Science Fiction writer and editor (3 Europa Awards for Best Author, Best Editor and Best Work of Fiction, 2 Urania Awards for Best Novel, 1 Golden Dragon Award for the promotion of International SF, 2 Italy Awards for Best Editor and Best Novel). He has published: *Antidoti umani*, *e-Doll*, *Nexhuman*, *Bloodbusters* and *I camminatori* (comprising *The Roamers* and *No/Mad/Land*). *Nexhuman* and *Bloodbusters* – translated in English by Sally McCorry and in Chinese by Zhang Fan and Hu Shaoyan – have been published in the US by Apex Books, in the UK by Luna press, and in China by Bofeng Culture.

Verso also works as editor and publisher of Future Fiction, a multicultural project dedicated to scouting and publishing the best World SF in translation from more than 40 countries and 14 languages, with authors like Ian McDonald, Ken Liu, Liu Cixin, Vandana Singh, Chen Qiufan, and others. Since 2019, he's the Honorary Director of the Fishing Fortress SF Academy of Chongqing. He may be found at futurefiction.org.

FLAME TREE PRESS
FICTION WITHOUT FRONTIERS
Award-Winning Authors & Original Voices

Flame Tree Press is the trade fiction imprint of Flame Tree Publishing, focusing on excellent writing in horror and the supernatural, crime and mystery, science fiction and fantasy. Our aim is to explore beyond the boundaries of the everyday, with tales from both award-winning authors and original voices.

•

•

Join our mailing list for free short stories, new release details, news about our authors and special promotions:

flametreepress.com